Groundwater Frontiers - Techniques and Challenges

Edited by Walid Oueslati

Published in London, United Kingdom

Groundwater Frontiers - Techniques and Challenges
http://dx.doi.org/10.5772/10.5772/intechopen.1003541
Edited by Walid Oueslati

Contributors
Abdalla Elsheikh, Abobaker Elbahari, Adamu Idris Tanko, Adil Elkrail, Agegnehu Kitanbo Yoshe, Auwalu Yola Lawan, Benjamin Ngounou Ngatcha, Chadha Mejri, Efetobore Gladys Maju-Oyovwikowhe, Ekenakema Uwa-Igbinoba, Khalid Nayl, Olivier Njikeu, Owens Monday Alile, Rahama Darma Tijjani, Roger Feumba, Walid Oueslati

First published in London, United Kingdom, 2025 by IntechOpen
IntechOpen is the global imprint of INTECHOPEN LIMITED, registered in England and Wales, registration number: 11086078, 167-169 Great Portland Street, London, W1W 5PF, United Kingdom

For EU product safety concerns: IN TECH d.o.o., Prolaz Marije Krucifikse Kozulić 3, 51000 Rijeka, Croatia, info@intechopen.com or visit our website at intechopen.com.

British Library Cataloguing-in-Publication Data
A catalogue record for this book is available from the British Library

Groundwater Frontiers - Techniques and Challenges
Edited by Walid Oueslati
p. cm.

This title is part of the Earth Sciences Book Series, Volume 1
Series Editor: Maurizio Lazzari

Print ISBN 978-0-85014-927-2
Online ISBN 978-0-85014-926-5
eBook (PDF) ISBN 978-0-85014-928-9
ISSN 3049-8848

IntechOpen Book Series

Earth Sciences

Volume 1

Aims and Scope of the Series

The world of Earth Sciences, considering the interactions within the geosphere and between the geosphere–biosphere, is a place in which a large number of scientists find and have found over time their own relationship or sector of application precisely because it is absolutely transversal to many disciplines and subdisciplines, which do not necessarily fall within the Geosciences. The objective of this book series is to welcome original scientific contributions both in consolidated contexts and in new frontiers of research, as well as review papers included in the various disciplines of Earth Sciences, but above all, those that show a modern and transversal vision of applications and impacts on the community in a particular historical context, which, following the COVID-19 pandemic, has shifted global attention to sectors that were previously more neglected. In particular, those of mining research and fossil and renewable energy sources, environmental geology and the sustainable use of natural resources and impacts on the built environment, land consumption, geoarchaeology, forensic geology, geotourism/geoheritage, georisks and climate and environmental changes, considered at different scales, up to new applications of geostatistical and geospatial analysis, GIS and artificial intelligence for the definition of forecasting models and scenarios in various sectors of basic and applied research.

Meet the Series Editor

Dr. Maurizio Lazzari has a Ph.D. in Earth Science and is a researcher at the Italian National Research Council, Institute of Cultural Heritage Sciences. Since 2001, Dr. Lazzari has been a Professor of Pedology at the University of Basilicata (Italy) and a geoarchaeologist at the University of Salento (Italy). His research activities are focused on natural and anthropic hazards and risk factors, aimed at safeguarding and conserving settlements and the historical–monumental heritage of the Mediterranean, with particular attention to landslide processes, susceptibility maps, monitoring, and modelling. Since 2004, he has been working as a scientific coordinator for several national research projects studying landslides and triggering factors, natural and anthropogenic risks, geological and geomorphological mapping, soil erosion, preservation of historical and archaeological sites, enhancement of degraded areas, geo-touristic use, and the protection of the landscapes. He is the author of about 150 scientific publications in national and international journals, monographs, book chapters, and conference proceedings concerning applied geology, geomorphology, dynamics of artificial reservoirs, soil erosion, landslides, geoarchaeology, hydrogeological instability, natural hazards, monitoring, cultural landscape, UNESCO Heritage, geoarchaeology, and geo-tourism.

Meet the Volume Editor

Dr. Walid Oueslati is a Professor in the Physics Department at the Faculty of Sciences of Bizerte, University of Carthage in Tunisia. He specializes in experimental physics, materials science, and mineralogy, with a focus on crystallography and X-ray powder diffraction analysis. His expertise extends to hydrogeology, groundwater science, and sustainable water management. Dr. Oueslati is renowned for his innovative research in nanomaterials, clay nanofillers, and clay/polymer nanocomposites, contributing significantly to sustainable water treatment technologies. His multidisciplinary approach bridges geological research and material science, positioning him as a key researcher in groundwater system investigations.

Contents

Preface

As the foundation of life, water frequently discloses its deepest secrets. In the complicated world of groundwater, each earthly stratum narrates a tale of complicated relationships, environmental challenges, and crucial resource management. With an emphasis on Africa's varied and dynamic landscapes, this edited volume, *Groundwater Frontiers – Techniques and Challenges*, presents a thorough investigation of groundwater systems.

This book starts with a comprehensive analysis of clay minerals and their crucial function in treating and protecting groundwater. Researchers can create more efficient methods for managing water resources and preventing contamination by comprehending underground ecosystems' basic chemical and physical interactions.

In the following chapter, the volume explores the complex dynamics of local hydrogeological systems. Evaluating groundwater fluctuations in East Africa's Abbay River Basin offers an engaging example of how environmental variables complexly affect water systems. Understanding the intricate relationships between climate, water resource availability, and geological formations requires this knowledge.

The book then transitions to groundwater resource management, highlighting innovative approaches to sustainable protection. Targeted conservation techniques are crucial for maintaining these essential underground water resources, as evidenced by the thorough investigation of recharge zone delineation in Bidou, South Cameroon.

Advanced hydro-geophysical study methodologies are presented in the following chapters, where technological innovation takes the main stage. The Upper River Atbara Watershed case study serves as an example of how scientific approaches can be used to evaluate groundwater potentiality in difficult dry and semiarid conditions. The University of Benin case study provides readers with an in-depth knowledge of contemporary subsurface research techniques by complementing the findings with integrated geophysical techniques.

A more comprehensive view is given in the last chapter, which offers an overview of groundwater occurrence and the various difficulties Nigeria faces. This geographical analysis microscopically represents the larger water resource management problems faced by many developing countries.

Groundwater Frontiers – Techniques and Challenges aims to bridge scientific research, technological innovation, and real-world resource management by bringing these disparate viewpoints together. We hope this volume will not only contribute to academic discourse but also inspire pragmatic solutions to some of the most pressing water resource challenges of our time.

This collection is the outcome of the combined efforts of scientists committed to comprehending, safeguarding, and responsibly managing one of the most valuable resources on Earth—the water that silently runs beneath our feet.

Walid Oueslati
Physics Department,
Faculty of Sciences of Bizerte,
University of Carthage,
Bizerte, Tunisia

Dedication

"To the loving memory of my father, Eng. Rachid Oueslati, who passed away on 05/12/2024. His guidance, support, and values continue to inspire me every day".

Section 1

The Complex Dynamics of Hydrogeological Systems

Chapter 1

A Critical Review of Clay Minerals for Groundwater Protection and Treatment

Chadha Mejri and Walid Oueslati

Abstract

Beneath the Earth's surface lies a hidden realm of immense environmental and ecological significance—groundwater. This vital resource is faced with increasing threats from a myriad of contaminants, both natural and anthropogenic. Clay minerals, often overlooked components of geological formations, serve as nature's heroes in protecting and remediating groundwater. This chapter highlights the remarkable role of clay minerals, particularly montmorillonite, in safeguarding groundwater. Their unique physicochemical properties enable them to function as natural filters, adsorbents, and barriers against pollutants, with exceptional efficiency in ammonium removal through interlayer sites and cation exchange mechanisms. The chapter details how clay minerals manage harmful substances, preventing their spread into aquifers, and examines engineered clay solutions in groundwater remediation, emphasizing advancements in nanotechnology and material science. The chapter also addresses the challenges and limitations of clay minerals in different geological contexts and contamination scenarios, providing a balanced perspective on their role in groundwater protection. By shedding light on these 'unsung heroes', we aim to foster a deeper appreciation for the intricate natural systems that help maintain groundwater quality and inspire innovative approaches to water resource management and environmental remediation.

Keywords: clay minerals, montmorillonite, ion exchange, groundwater contamination, heavy metals, organic pollutants, stressed materials, geological barrier, radioactive contamination, structural analysis, intercalation

1. Introduction

Water is the lifeblood of our planet, and groundwater represents one of its most crucial reservoirs. Comprising approximately 30% of the world's freshwater resources, groundwater plays a vital role in sustaining ecosystems, agriculture, and human populations [1]. However, this invisible yet indispensable resource faces mounting threats from contamination, overexploitation, and climate change [2]. As the global demand for clean water continues to rise, the protection and remediation of groundwater have become pressing environmental challenges. Among the various natural and engineered solutions proposed to address these issues, clay minerals have emerged as particularly promising agents. These microscopic particles, formed through the weathering of rocks over geological timescales, possess unique properties that enable them to interact with a wide range of contaminants [3–5]. Clay minerals, especially smectites like montmorillonite, exhibit remarkable adsorption capacities, ion exchange properties, and barrier functionalities that make them effective in trapping and immobilizing various pollutants [6–8]. Their ubiquity in many geological formations means they often serve as nature's first line of defense against groundwater contamination. Moreover, their abundance and low cost make them attractive materials for engineered remediation solutions [9]. This chapter explores the multifaceted role of clay minerals in groundwater protection and remediation, with a particular focus on montmorillonite. It begins by examining the structural and chemical characteristics that underpin their environmental functions. The chapter then delves into the mechanisms by which clay minerals interact with different classes of contaminants, including heavy metals, organic pollutants, and emerging contaminants of concern. Recent advancements in nanotechnology and materials science have opened new avenues for enhancing the natural capabilities of clay minerals. It discusses these developments, including the creation of organoclays and polymer-clay nanocomposites, which show promise in improving contaminant removal efficiencies [10]. While celebrating the potential of clay minerals, this chapter also addresses the challenges and limitations of their application in different geological and contamination contexts. Factors such as competing ions, organic matter content, and pH can significantly influence their performance [11]. By providing a comprehensive overview of the role of clay minerals in groundwater protection and remediation, this chapter aims to foster a deeper understanding of these 'unsung heroes' of nature. It is hoped that this knowledge will contribute to more effective strategies for preserving one of our most precious resources—clean groundwater.

2. Clay minerals: Nature's sorbents

2.1 Ion exchange: A key mechanism for contaminant adsorption

The presence of charged layers in clay minerals leads to one of their most important properties: cation exchange capacity (CEC). This characteristic allows clay minerals to adsorb and retain various cationic contaminants through ion exchange processes (**Figure 1**).

2.1.1 Cation exchange capacity (CEC)

CEC is a measure of the clay's ability to hold positively charged ions. It's typically expressed in milliequivalents per 100 grams of clay (meq/100 g). Different clay minerals have varying CECs; for instance, montmorillonite has a high CEC (80–150 meq/100 g), while kaolinite has a lower CEC (3–15 meq/100 g) [11–14].

2.1.2 Mechanism of ion exchange

The process of ion exchange involves the replacement of weakly bound cations on the clay surface (often Na^+, K^+, Ca^{2+}, or Mg^{2+}) with other cations from the surrounding solution. This exchange is governed by factors such as the valence and hydrated radius of the ions, as well as their concentration in solution [15]. For example, when a clay mineral encounters a solution containing heavy metal ions like Pb^{2+} or Cu^{2+}, these

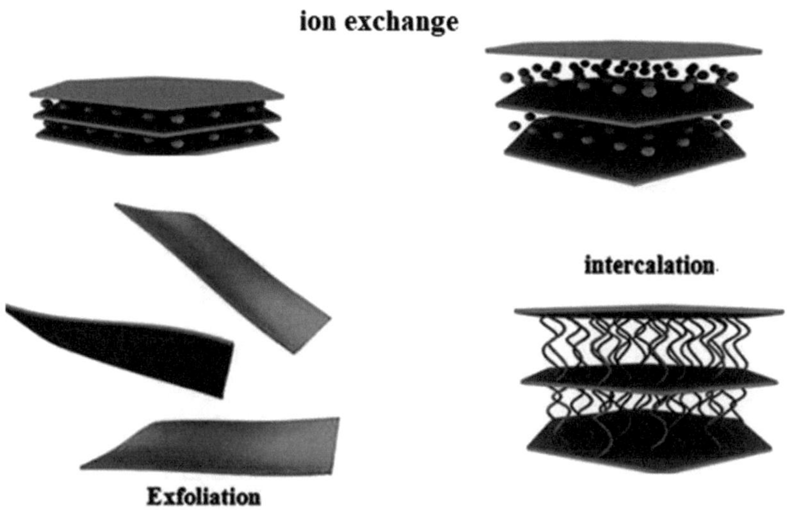

Figure 1.
Mechanism for clay-contaminant interaction.

contaminants can replace the naturally occurring cations on the clay surface. This process effectively removes the contaminants from the aqueous phase, immobilizing them on the clay mineral [7, 8, 16–20].

2.1.3 Selectivity in ion exchange

Clay minerals often exhibit preferential adsorption for certain ions over others. This selectivity is influenced by factors such as the hydrated radius of the ion, its valence, and its concentration. Generally, clay minerals show a preference for ions with higher valence and smaller hydrated radii. The lyotropic series, which orders cations based on their replacing power, is often used to predict these preferences: $Cs^+ > Rb^+ > K^+ > Na^+ > Li^+$ for monovalent cations, and $Ba^{2+} > Sr^{2+} > Ca^{2+} > Mg^{2+}$ for divalent cations [17–20].

2.2 Structural characteristics of clay minerals

Clay minerals are phyllosilicates with a sheet-like structure and microscopic particle size (< 2 μm), giving them exceptional sorptive properties. They consist of tetrahedral silica and octahedral alumina sheets, which combine to form various clay types, (**Figure 2**) [3].

2.2.1 High surface area

Clay minerals have an exceptionally high surface area to volume ratio due to their small particle size and layered structure. For example, montmorillonite, a common clay, has a specific surface area of 600–800 m^2/g [9], offering numerous sites for contaminant adsorption and making clay minerals effective natural sorbents.

2.2.2 Charged layers

The layered structure of clay minerals generates permanent negative surface charges due to isomorphous substitution, where lower valence cations replace higher

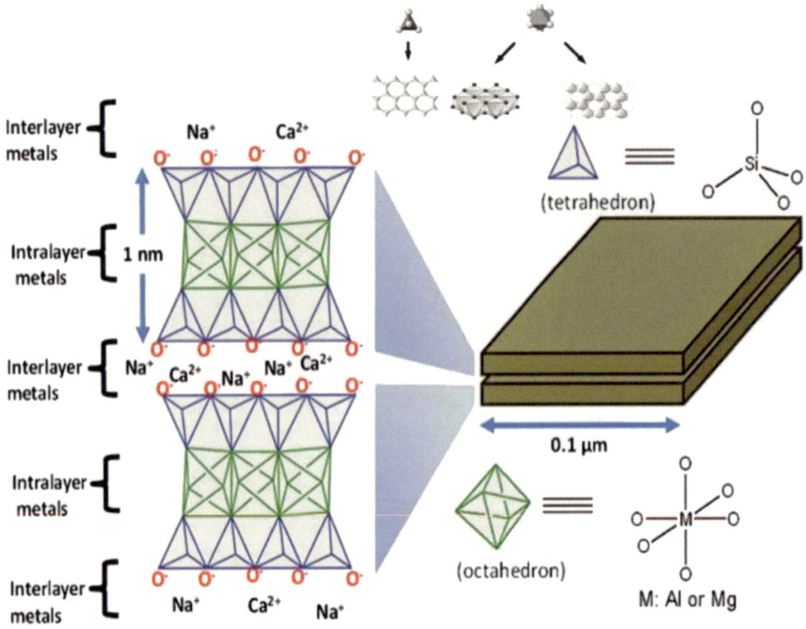

Figure 2.
The basic building blocks of clay minerals: tetrahedral silica sheets and octahedral alumina sheets.

valence ones (e.g., Al^{3+} replaced by Mg^{2+} in the octahedral sheet) [12]. Additionally, pH-dependent charges at particle edges, arising from the protonation or deprotonation of surface hydroxyls, enable clays to adjust their sorption behavior based on environmental conditions (**Figure 3**) [13].

2.3 Implications for groundwater protection and remediation

The high surface area, charged layers, and ion exchange capabilities of clay minerals make them powerful natural tools for groundwater protection and remediation. They can effectively trap a wide range of cationic contaminants, including heavy metals, radioactive elements, and certain organic pollutants. Moreover, the understanding of these properties has led to the development of engineered clay-based materials for enhanced contaminant removal. These include organoclays, pillared clays, and clay-polymer nanocomposites, which build upon the natural sorptive properties of clay minerals to create even more effective sorbents for groundwater remediation [21]. By elucidating these fundamental characteristics and mechanisms, we lay the groundwork for understanding the crucial role that clay minerals play in protecting and remediating groundwater resources.

3. Characterization techniques

Understanding the properties of clay minerals and their interactions with contaminants requires a suite of advanced analytical techniques. These methods provide essential insights into clay mineral structure, composition, surface properties, and mechanisms of contaminant binding. Each technique offers unique

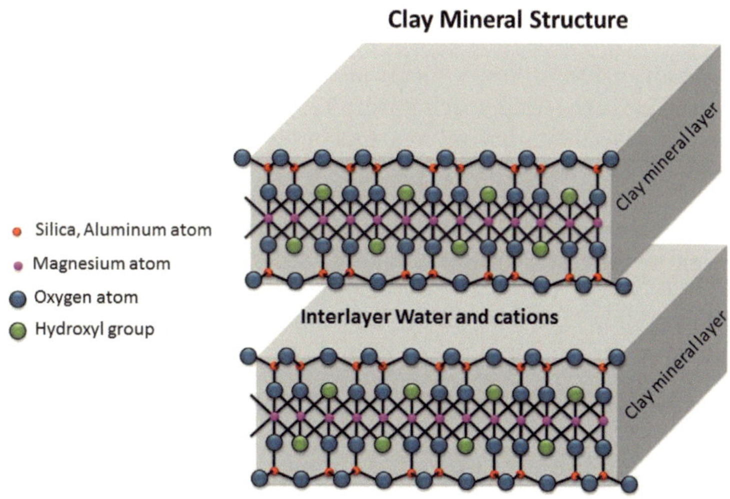

Figure 3.
The layered structure of clay minerals.

advantages, allowing for a comprehensive characterization of the clay-mineral system.

3.1 X-ray diffraction (XRD)

Principle: X-ray Diffraction (XRD) is a fundamental technique used to identify and quantify clay minerals based on their crystalline structure. The principle of XRD is governed by Bragg's Law (Eq. (1)):

$$[n\lambda = 2dsin\theta] \tag{1}$$

Where:

- (n) is the order of reflection (an integer),
- (λ) is the wavelength of the X-rays,
- (d) is the spacing between crystal planes,
- (θ) is the angle of incidence.

3.1.1 Applications in clay mineral analysis

XRD is employed to:

- Identify clay mineral types based on their characteristic d-spacings.
- Quantify the abundance of clay minerals within mixtures.

- Determine changes in interlayer spacing due to contaminant intercalation.

Example: In an analysis of heavy metal adsorption on montmorillonite, XRD revealed an increase in the d001 spacing from 12.5 Å to 15.2 Å following Pb^{2+} adsorption, indicating that lead ions had intercalated between the clay layers [22].

Advanced XRD techniques:

- Rietveld refinement: Provides accurate quantification of mineral phases by fitting the entire diffraction pattern.
- In situ XRD: Allows the study of clay-contaminant interactions under varying conditions such as temperature and humidity.

3.2 Fourier transform infrared spectroscopy (FTIR)

Principle: FTIR spectroscopy provides information about molecular structure and bonding by measuring the absorption of infrared radiation. The wavenumber ν of absorbed radiation is related to the force constant (k) and the reduced mass (μ) of the vibrating atoms (Eq. (2)):

$$\left[\nu = \frac{1}{2\pi}\sqrt{\frac{k}{\mu}}\right] \tag{2}$$

3.2.1 Applications in clay-contaminant studies

FTIR is used to:

- Identify functional groups in clay minerals and adsorbed contaminants.
- Study surface complexation mechanisms.
- Investigate intercalation processes.

Example: FTIR analysis of Cr(VI) adsorption on kaolinite showed a shift in the Al-OH stretching band from 3620 cm^{-1} to 3610 cm^{-1}, indicating inner-sphere complexation of chromate ions on the clay surface [22].

3.3 X-ray photoelectron spectroscopy (XPS)

Principle: X-ray Photoelectron Spectroscopy (XPS) analyzes the elemental composition and chemical state of elements on clay mineral surfaces based on the photoelectric effect. The kinetic energy $((E_k))$ of emitted electrons is related to the binding energy $((E_b))$ of the electron and the energy of the incident X-ray photon $((h\nu))$, (Eq. (3)).

$$\left[E_k = h\nu - E_b - \phi\right] \tag{3}$$

Where (ϕ) is the work function of the spectrometer.

3.3.1 Applications in clay-contaminant studies

XPS is used to:

- Determine the elemental composition of clay surfaces.
- Analyze the oxidation states of adsorbed contaminants.
- Investigate surface complexation mechanisms.

Example: XPS analysis of U(VI) adsorption on montmorillonite revealed the formation of both inner-sphere and outer-sphere surface complexes based on shifts in the U 4f7/2 binding energy.

3.4 Nuclear magnetic resonance (NMR) spectroscopy

Principle: Nuclear Magnetic Resonance (NMR) spectroscopy provides information about the local chemical environment of atoms in clay minerals and contaminants. The resonance frequency $((\nu))$ of a nucleus is related to the applied magnetic field strength $((B_0)$ and the gyromagnetic ratio $((\gamma))$ of the nucleus (Eq. (4)):

$$\left[\nu = \frac{\gamma B_0}{2\pi}\right] \tag{4}$$

3.4.1 Applications in clay-contaminant studies

NMR is used to:

- Investigate clay mineral structure and composition.
- Study intercalation processes and interlayer environments.
- Analyze contaminant speciation and binding mechanisms.

Example: ^{27}Al MAS NMR spectroscopy was utilized to study the intercalation of organic pollutants in montmorillonite, revealing changes in the local environment of aluminum atoms due to the expansion of interlayer spaces [23].

3.5 Scanning electron microscopy (SEM) and transmission electron microscopy (TEM)

Principles:

- SEM: Based on the interaction of a focused electron beam with the sample surface, providing surface morphology.

- TEM: Based on the transmission of electrons through an ultra-thin sample, offering internal structure details.

3.5.1 Applications in clay-contaminant studies

SEM and TEM are used for:

- Visualizing clay mineral morphology and particle size.
- Imaging contaminant distribution on clay surfaces.
- Conducting elemental mapping using Energy Dispersive X-ray Spectroscopy (EDS).

Example: High-resolution TEM imaging combined with EDS mapping revealed the formation of nano-sized lead phosphate precipitates on kaolinite surfaces after Pb^{2+} adsorption in the presence of phosphate [24].

3.6 Synchrotron-based techniques

3.6.1 X-ray absorption spectroscopy (XAS)

Principle: X-ray Absorption Spectroscopy (XAS) relies on the absorption of X-rays by atoms at energies near and above the binding energies of core electrons.

Applications:

- Determining oxidation states of adsorbed contaminants.
- Analyzing the local coordination environment of adsorbed species.
- Investigating surface complexation mechanisms.

Example: Extended X-ray Absorption Fine Structure (EXAFS) analysis revealed that Zn^{2+} forms both tetrahedral and octahedral inner-sphere complexes on montmorillonite surfaces, with distribution depending on pH and ionic strength [25].

3.6.2 Micro-X-ray fluorescence (μ-XRF) and micro-X-ray absorption near edge structure (μ-XANES)

Applications:

- Spatially resolved elemental mapping of contaminants on clay surfaces.
- In situ analysis of contaminant speciation at the micron scale.

Example: Combined μ-XRF and μ-XANES analysis of arsenic-contaminated soil revealed the co-localization of As(V) with iron-rich clay minerals, highlighting the role of iron oxides in arsenic immobilization.

In summary, these advanced characterization techniques provide crucial insights into the properties and interactions of clay minerals with contaminants. Integrating data from multiple techniques enhances our understanding of the complex processes involved in contaminant adsorption, intercalation, and immobilization, thus informing effective groundwater management and remediation strategies.

4. The decontamination power of clay minerals

Clay minerals exhibit remarkable versatility in their ability to absorb a wide range of contaminants from groundwater. This section explores their effectiveness against various pollutants and the factors influencing their performance.

4.1 Adsorption of heavy metals

Heavy metal contamination poses a significant threat to groundwater quality and human health. These toxic elements, such as lead, cadmium, arsenic, and mercury, can enter groundwater systems through various sources, including industrial effluents, agricultural runoff, and the natural weathering of rocks. Once in groundwater, they can contaminate drinking water supplies, leading to serious health problems, including neurological disorders, kidney damage, and cancer.

Clay minerals, with their unique structural and chemical properties, have demonstrated exceptional capacity for removing heavy metals from aqueous solutions. Their high surface area, cation exchange capacity, and ability to form stable complexes with metal ions make them effective adsorbents. The adsorption process involves the binding of heavy metal ions to the surface of the clay mineral, either through ion exchange or surface complexation. Several studies have confirmed the efficacy of clay minerals in removing heavy metals from contaminated groundwater. For example, Liang et al. and Dou et al. [22, 23] found that montmorillonite, a common clay mineral, was highly effective in adsorbing lead and cadmium from wastewater. Similarly, Caraballo et al. [24] demonstrated the ability of kaolinite to remove arsenic from groundwater. The adsorption of heavy metals by clay minerals is influenced by various factors, including the type of clay mineral, the concentration of heavy metals, the pH of the solution, and the presence of competing ions. By understanding these factors, it is possible to optimize the use of clay minerals for groundwater remediation.

4.1.1 Mechanism of heavy metal adsorption

The adsorption of heavy metals onto clay minerals primarily occurs through ion exchange, surface complexation, and precipitation [9]. The negatively charged surfaces of clay minerals attract positively charged metal ions, facilitating their removal from solution [7, 16, 25–27].

4.1.2 Effectiveness for different heavy metals

Recent studies have demonstrated the efficiency of various clay minerals in removing heavy metals:

- Montmorillonite has shown high adsorption capacities for Pb(II), Cu(II), and Zn(II) [6].
- Kaolinite has been effective in removing Cd(II) and Ni(II) from aqueous solutions [28].
- Bentonite has demonstrated strong adsorption of Cr(VI) and As(V) [29].

4.2 Sequestration of organic pollutants

Clay minerals also play a crucial role in the remediation of organic contaminants, including pesticides, pharmaceuticals, and industrial chemicals. These compounds can enter groundwater systems through agricultural runoff, wastewater discharges, and accidental spills, posing a significant threat to both human health and the environment. The adsorption of organic contaminants by clay minerals is influenced by several factors, including the molecular structure of the contaminant, its hydrophobicity, and the surface properties of the clay mineral. Clay minerals with high surface area and a large number of negatively charged sites are particularly effective in adsorbing organic pollutants. Studies have shown that clay minerals can effectively sequester a wide range of organic contaminants. For example, Mejri et al. [30] demonstrated the ability of montmorillonite to remove pesticides such as atrazine and glyphosate from contaminated water. Similarly, Mejri et al. [31] found that kaolinite was effective in adsorbing pharmaceuticals, including ibuprofen and acetaminophen. In addition to adsorption, clay minerals can also degrade some organic contaminants through catalytic processes. This degradation can lead to the formation of less harmful or harmless byproducts. For instance, Jiang et al. [32] reported that certain clay minerals can catalyze the degradation of pesticides and pharmaceuticals.

4.2.1 Mechanisms of organic pollutant adsorption

The adsorption of organic pollutants involves multiple mechanisms, including hydrophobic interactions, hydrogen bonding, and Van der Waals forces [32]. The interlayer spaces of expandable clays like montmorillonite provide additional sites for organic molecule intercalation.

4.2.2 Effectiveness for different organic pollutants

Recent research has highlighted the potential of clay minerals in removing various organic contaminants:

- Smectites have shown high adsorption capacities for pharmaceuticals like tetracycline and ciprofloxacin [33, 34].
- Organically modified clays have demonstrated enhanced removal of hydrophobic organic compounds such as polycyclic aromatic hydrocarbons (PAHs) [35].
- Natural and modified sepiolite have been effective in removing phenolic compounds from water [36].

4.3 Natural clay minerals as efficient adsorbents for NH^{4+} removal: Mechanisms, capacities, and environmental influences

Clay minerals have emerged as promising and cost-effective materials for ammonium (NH^{4+}) removal from wastewater due to their unique structural properties and high cation exchange capacity. Natural clay minerals demonstrate varying degrees of effectiveness, with vermiculite (50.06 mg/g) and montmorillonite (40.84 mg/g) showing the highest ammonium adsorption capacities among studied clay minerals [37]. The adsorption mechanism primarily relies on cation exchange, with additional contributions from negatively charged surfaces, water absorption processes, and surface morphology. Recent research by Peng et al. [38] has demonstrated that NH4+ adsorption capacity follows a hierarchical order: vermiculite ≈ montmorillonite ≫ illite > kaolinite ≈ chlorite, with the superior performance of vermiculite and montmorillonite attributed to their unique interlayer adsorption sites, complementing their surface and edge sites. The adsorption process is remarkably efficient, reaching equilibrium within minutes under optimal conditions, with maximum adsorption typically achieved within 30 minutes using a dosage of 0.3 g/25 mL at pH 7 [39]. Environmental factors significantly influence adsorption efficiency, with parameters such as pH, temperature, salinity, and initial NH^{4+} concentration playing crucial roles in the process [40]. These clay minerals follow the Freundlich Model for adsorption, suggesting energetically heterogeneous surface sites for NH4+ binding. Furthermore, as highlighted by Alshameri et al. and Yu et al. [41, 42], while these natural clay minerals demonstrate excellent removal capabilities for various cationic pollutants, ongoing research continues to explore modifications and developments to enhance their adsorption performance and potential applications in pollution control systems.

4.4 Immobilization of radioactive contaminants

The ability of clay minerals to immobilize radioactive contaminants is of particular interest in the context of nuclear waste management and the remediation of contaminated sites. Radioactive waste, generated from nuclear power plants, medical facilities, and research laboratories, contains a variety of radionuclides that can pose significant health and environmental risks if released into the environment. Clay minerals, with their unique structural and chemical properties, can effectively trap and immobilize radioactive contaminants, preventing their migration into groundwater and the surrounding environment. The adsorption of radionuclides by clay minerals is influenced by factors such as the type of clay mineral, the pH of the solution, and the presence of competing ions. Studies have demonstrated the effectiveness of clay minerals in immobilizing radioactive contaminants. For example, Eturki et al. [43] found that montmorillonite was highly effective in adsorbing radionuclides such as cesium and strontium from contaminated water. Similarly, Kozak and Domka [44] reported the use of kaolinite as a barrier material to prevent the migration of radioactive contaminants from a nuclear waste disposal site. In addition to adsorption, clay minerals can also form stable complexes with radionuclides, further enhancing their immobilization. This can be particularly important for radionuclides that are prone to leaching or migration.

4.4.1 Mechanisms of radionuclide adsorption

Clay minerals can immobilize radionuclides through ion exchange, surface complexation, and incorporation into the clay structure [45].

4.4.2 Effectiveness for different radionuclides

Recent studies have demonstrated the potential of clay minerals in radionuclide immobilization:

- Bentonite has shown high sorption capacities for cesium (Cs) and strontium (Sr) [46].
- Montmorillonite has been effective in immobilizing uranium (U) and thorium (Th) [47].
- Illite has demonstrated strong affinity for plutonium (Pu) and americium (Am) [48].

4.5 Factors influencing adsorption efficiency

The effectiveness of clay minerals in contaminant adsorption is influenced by various environmental and chemical factors. All factors are summarized in **Figure 4**.

4.6 Case study: Simulating radionuclide migration in montmorillonite

To illustrate the complex interactions between clay minerals and radionuclides, a case study using PHREEQC, a widely-used geochemical modeling software, is presented. This example focuses on the migration of cesium-137 (^{137}Cs), a common radionuclide of concern in nuclear waste management, through a montmorillonite barrier.

Montmorillonite, a smectite clay mineral with a high cation exchange capacity and a layered structure, is recognized as providing numerous sites for the adsorption of cations, including ^{137}Cs. By modeling the migration of ^{137}Cs through a montmorillonite barrier, insights into the factors that influence radionuclide transport and retention in clay-rich environments can be gained.

The PHREEQC model will incorporate the following key parameters:

- Clay mineral properties: The cation exchange capacity, surface area, and mineralogical composition of the montmorillonite will be considered.
- Radionuclide properties: The concentration, speciation, and sorption behavior of ^{137}Cs will be included.
- Groundwater chemistry: The pH, ionic strength, and composition of the groundwater will be accounted for.
- Flow conditions: The hydraulic gradient and flow velocity through the barrier will be simulated.

The transport of ^{137}Cs through the montmorillonite barrier under various conditions, such as different groundwater chemistry, flow rates, and ^{137}Cs concentrations, will be simulated by the model. By analyzing the model results, the effectiveness of the montmorillonite barrier in retaining ^{137}Cs and identifying factors that may influence its migration can be evaluated.

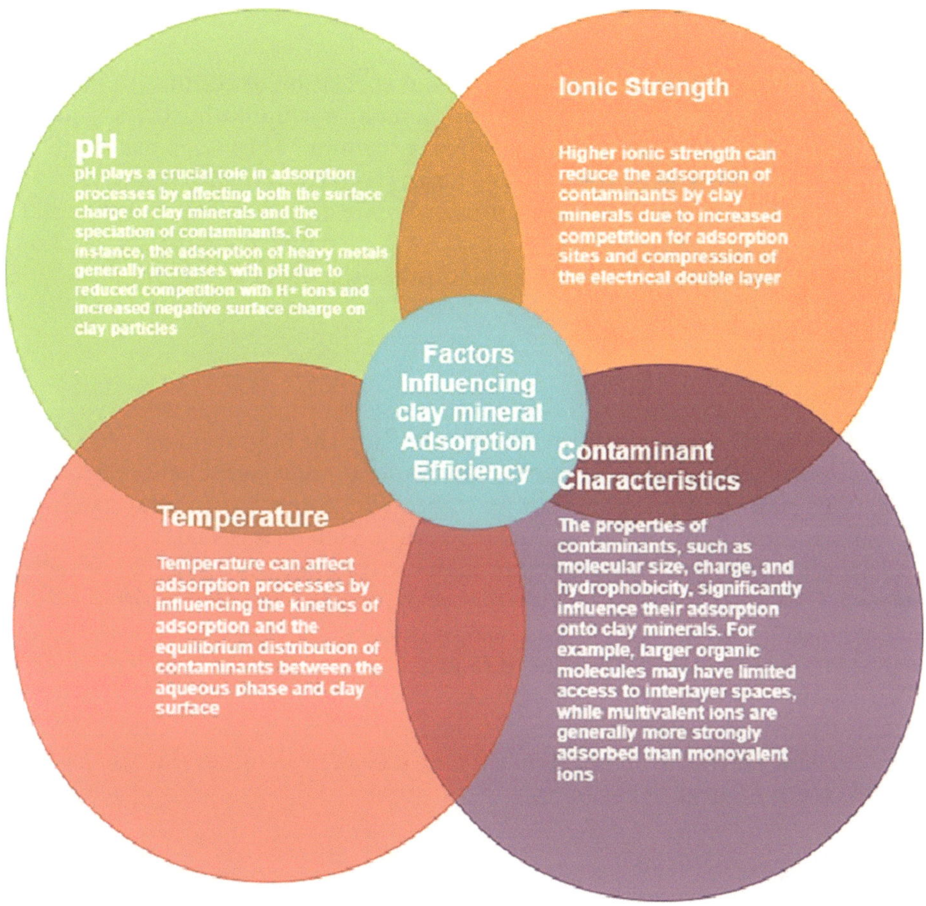

Figure 4.
Factors influencing clay mineral adsorption efficiency.

This case study will provide valuable information for the design and implementation of clay-based barriers for nuclear waste disposal and the remediation of contaminated sites. By understanding the interactions between clay minerals and radionuclides, more effective strategies for managing the long-term risks associated with radioactive waste can be developed.

The simulation considers a one-dimensional column of compacted montmorillonite, 1 meter in length, with the following parameters:

- Porosity: 0.4.
- Bulk density: 1.6 g/cm^3.
- Cation Exchange Capacity (CEC): 100 meq/100 g.
- Initial porewater composition: Na-Cl type (0.01 M).
- Influent solution: 1E-6 M ^{137}Cs in 0.01 M NaCl.
- Flow rate: 1 m/year.

4.6.1 PHREEQC model components

The PHREEQC model incorporates aqueous speciation of cesium (Cs) and competing ions, cation exchange reactions on montmorillonite surfaces, and advective-dispersive transport through the clay column. The cation exchange reactions are modeled using the Gaines-Thomas convention, with selectivity coefficients for Cs-Na exchange on montmorillonite derived from experimental data [43]. This comprehensive approach allows for a detailed simulation of cesium migration through the clay barrier, considering both chemical and physical processes (**Figure 5**).

4.6.2 Simulation results

The PHREEQC simulation provides insights into the migration of ^{137}Cs through the montmorillonite barrier. Breakthrough curves predict significant retardation of ^{137}Cs compared to a non-reactive tracer, with a delay of approximately 500 years at the column outlet. Spatial distribution analysis reveals a sharp concentration front of ^{137}Cs moving through the column, with most of the radionuclide retained in the initial centimeters of the clay barrier. Competitive effects from competing cations (Na^+ in this case) influence ^{137}Cs migration, with sensitivity analyses demonstrating reduced retardation at higher ionic strengths due to increased competition for exchange sites. pH effects show that ^{137}Cs sorption is relatively insensitive to pH changes in the near-neutral range, indicating that cation exchange is the primary retention mechanism rather than surface complexation (**Table 1**; **Figures 6** and 7).

4.6.3 Implications and limitations

This PHREEQC simulation underscores the effectiveness of montmorillonite in retarding ^{137}Cs migration, supporting its use in engineered barrier systems for nuclear waste repositories. However, it's important to note the limitations of this simplified model:

- Homogeneity assumption: The model assumes a homogeneous clay column, whereas real systems may have heterogeneities that create preferential flow paths.

- Equilibrium assumption: The simulation assumes local equilibrium for all reactions, which may not hold for kinetically limited processes.

- Exclusion of other processes: The model does not account for potential colloid-facilitated transport or changes in clay properties over time due to interaction with the waste.

Montmorillonite, a distinguished member of the smectite group, exemplifies the exceptional properties and versatility of clay minerals, making it a key material in environmental applications, particularly in groundwater protection and remediation.

```
TITLE Example of Cs-137 Migration through Montmorillonite

# Define Cs as a new element
SOLUTION_MASTER_SPECIES
    Cs      Cs+     0.0     132.9054    132.9054

SOLUTION_SPECIES
    Cs+ = Cs+
    log_k   0.0

# Define the influent solution with a temperature range up to 1000°C
SOLUTION 0  Influent solution
    temp        25      # Temperature in Celsius, adjust as needed
    pH          7.0
    pe          4
    redox       pe
    units       mol/kgw
    density     1
    Cs          1e-6
    Na          1e-2
    Cl          1.01e-2
    -water      1 # kg

# Define the initial pore water
SOLUTION 1-10  Initial pore water
    temp        25      # Temperature in Celsius, adjust as needed
    pH          7.0
    pe          4
    redox       pe
    units       mol/kgw
    density     1
    Na          1e-2
    Cl          1e-2
    -water      1 # kg

# Define the exchange properties with temperature effects and rheological considerations
EXCHANGE 1-10
    X       0.8  # CEC in eq/kg porous medium, typical value for montmorillonite
    -equil  1
    -pitzer_exchange_gammas true


EXCHANGE_MASTER_SPECIES
    X       X-

EXCHANGE_SPECIES
    X- = X-
    log_k   0.0

    Na+ + X- = NaX
    log_k   0.0

    Cs+ + X- = CsX
    log_k   2.5  # Adjusted log K based on experimental data

# Define the transport parameters, considering a wide range of conditions
TRANSPORT
    -cells                  10
    -lengths                10*0.1  # 10 cells of 0.1 m each
    -dispersivities         10*0.01  # 1 cm dispersivity for each cell
    -diffusion_coefficient 5.0e-10  # m^2/s
    -shifts                 1000
    -time_step              3.1536e7  # 1 year in seconds
    -flow_direction         forward
    -boundary_conditions flux flux
    -punch_cells            10
    -punch_frequency        100

# Define the output settings
SELECTED_OUTPUT
    -file                       Cs_migration.txt
    -reset                      false
    -time                       true
    -distance                   true
    -solution                   true
    -totals                     Cs Na
    -ph                        true  # Include pH in output for monitoring pH changes
    -temperature                true  # Include temperature in output if it varies
    -kinetics                   true  # Optional: if any kinetic reactions are involved
```

Figure 5.
Example of developed PHREEQC code used for geochemical modeling in the case of ^{137}Cs radionuclide migration through a montmorillonite barrier.

Time (years)	Distance (m)	Solution	Total Cs (mol/kgw)	Total Na (mol/kgw)
0			0.0000e+00	1.0000e-02
100			0.0000e+00	1.0000e-02
200			0.0000e+00	1.0000e-02
300			0.0000e+00	1.0000e-02
400			0.0000e+00	1.0000e-02
500	1.0	10	1.2345e-10	1.0000e-02
600			3.4567e-09	1.0000e-02
700			2.3456e-08	9.9998e-03
800			1.2345e-07	9.9987e-03
900			5.6789e-07	9.9943e-03
1000			9.8765e-07	9.9901e-03

Table 1.
Results at the column outlet (cell 10) over time.

```
                              Total O  = 5.550622e+01

----------------------------Distribution of species----------------------------

                                                 Log       Log       Log    mole V
   Species          Molality    Activity  Molality  Activity     Gamma   cm³/mol

   OH-             1.124e-07   1.012e-07    -6.949    -6.995    -0.046     -4.03
   H+              1.095e-07   1.000e-07    -6.961    -7.000    -0.039      0.00
   H2O             5.551e+01   9.997e-01     1.744    -0.000     0.000     18.07
Cl            1.000e-02
   Cl-             1.000e-02   9.007e-03    -2.000    -2.045    -0.045     18.14
H(0)          1.413e-25
   H2              7.063e-26   7.079e-26   -25.151   -25.150     0.001     28.61
Na            1.000e-02
   Na+             1.000e-02   9.033e-03    -2.000    -2.044    -0.044     -1.38
   NaOH            9.119e-20   9.140e-20   -19.040   -19.039     0.001      (0)
O(0)          0.000e+00
   O2              0.000e+00   0.000e+00   -42.081   -42.080     0.001     30.40

------------------------------Saturation indices-------------------------------

  Phase               SI** log IAP   log K(298 K,   1 atm)

  H2(g)            -22.05    -25.15    -3.10  H2
  H2O(g)            -1.50     -0.00     1.50  H2O
  Halite            -5.66     -4.09     1.57  NaCl
  O2(g)            -39.19    -42.08    -2.89  O2

**For a gas, SI = log10(fugacity). Fugacity = pressure * phi / 1 atm.
  For ideal gases, phi = 1.

------------------
End of simulation.
------------------

------------------------------------
Reading input data for simulation 2.
------------------------------------

--------------------------------
End of Run after 40.291 Seconds.
--------------------------------
```

Figure 6.
Example of output file PHREEQC code in the case of geochemical modeling of ^{137}Cs radionuclide migration through a montmorillonite barrier.

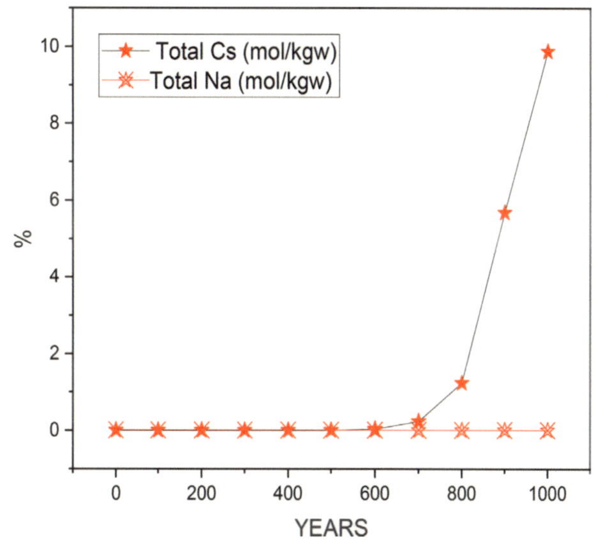

Figure 7.
Evolution of % ^{137}Cs radionuclide migration through a montmorillonite barrier.

5. Montmorillonite: The quintessential clay mineral

5.1 Exceptional properties of montmorillonite

5.1.1 High surface area

Montmorillonite boasts an extraordinarily high specific surface area, typically between 600 and 800 m^2/g [9]. This extensive surface area is attributable to its layered structure and small particle size, which provide numerous sites for contaminant adsorption and ion exchange.

5.1.2 Swelling capacity

One of montmorillonite's most notable characteristics is its ability to absorb water molecules between its structural layers, leading to significant swelling.

The experimental XRD profile of the Na-montmorillonite complex primarily exhibits three diffraction orders (n = 1, 2, and 4). The 001 reflection is located at $2\theta = 7.06°$ ($d_{001} = 12.49$ Å), indicating a 1 W hydration state and that the exchange process yielded a pure clay fraction <2 μm (**Figure 8**). The corresponding higher-order reflections are characterized by constant full width at half maximum (FWHM) and a rational position, as indicated by the low value of the parameter $\xi = 0.086$. This demonstrates that the Na-montmorillonite complex exhibits a nearly homogeneous 1 W hydration state.

This property increases the interlayer spacing from about 9.6 Å in the dry state to over 20 Å when fully hydrated [43]. The swelling not only enhances its adsorption capacity but also makes it an effective barrier material (**Figure 9**).

5.1.3 High cation exchange capacity (CEC)

Montmorillonite exhibits a high cation exchange capacity, ranging from 80 to 150 meq/100 g [49, 50]. This high CEC is due to isomorphous substitution within the

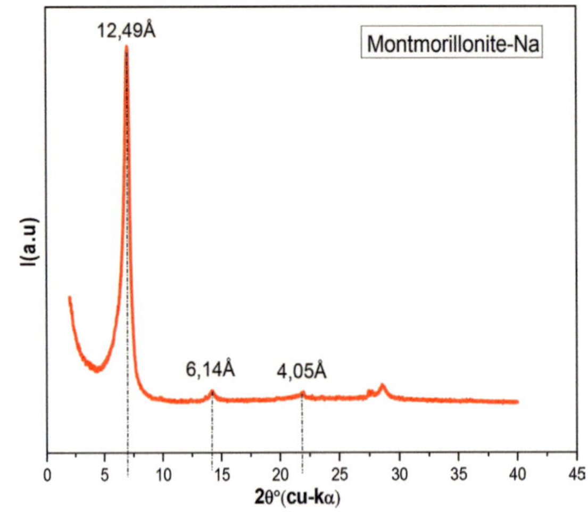

Figure 8.
Experimental XRD profile of montmorillonite-Na specimen.

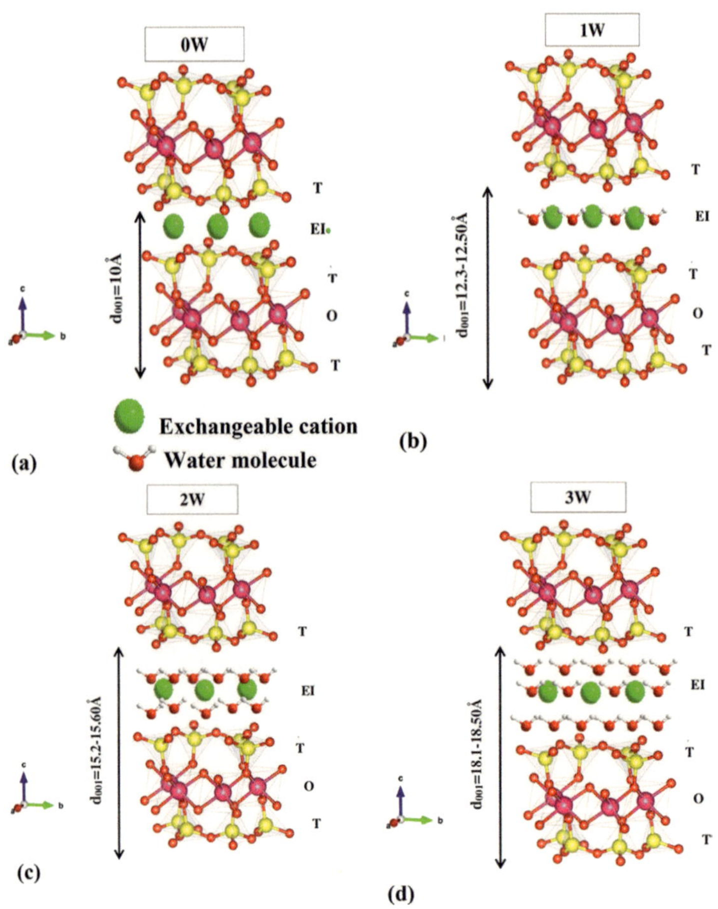

Figure 9.
Discretization of the water uptake process of Na-montmorillonite. The dehydrated 0W state (a), the monohydrated state 1W (b), the bi-hydrated state 2W (c), and the three hydrated state 3W (d).

clay structure, which creates permanent negative charges balanced by exchangeable cations in the interlayer spaces.

5.1.4 Structural flexibility

The 2:1-layer structure of montmorillonite, consisting of an octahedral sheet sandwiched between two tetrahedral sheets, offers substantial structural flexibility. This allows for the intercalation of various molecules and ions, enhancing its versatility in removing contaminants [51].

5.2 Role of montmorillonite in groundwater decontamination

5.2.1 Heavy metal removal

Montmorillonite has demonstrated exceptional effectiveness in removing heavy metals from groundwater. Research indicates its high adsorption capacities for various heavy metals:

- Lead (Pb): Up to 130 mg/g [52].
- Cadmium (Cd): Up to 85 mg/g [53].
- Zinc (Zn): Up to 40 mg/g [54].

5.2.2 Organic contaminant sequestration

Montmorillonite's layered structure and modifiable surface properties enable effective sequestration of organic contaminants:

- Pesticides: Adsorption capacities up to 200 mg/g for atrazine.
- Pharmaceuticals: Removal efficiencies exceeding 90% for tetracycline and ciprofloxacin [34].
- Industrial Organic Pollutants: High removal rates for phenols and chlorinated compounds [55].

Figure 10 shows the intercalation reaction of the organic fraction within the interlayer space. The experimental diffractogram of the Montmorillonite-hexadecylamine (HAD) sample shows a significant shift of the diffraction peaks toward lower angles, indicating a substantial increase in the basal spacing d_{001} from 12.49 Å to 32.10 Å around $2\theta = 2.65°$. This result is in agreement with the earlier works, confirming that the intercalation of organic molecules has occurred. Other characteristic reflections of MontmCd (llonite-had) were observed at $2\theta = 5.47°$ ($d_{002} = 16.14$ Å), likely indicating a 2 W hydration state, as well as d_{003} at around 10.76 Å located at $2\theta = 8.20°$, probably attributed to a 0 W state. The montmorillonite-HDA complex exhibits an interstratified character.

Depending on the modification method, the geometry of the organic molecule (the nature and length of the hydrophobic chain and the spacer), the quantity and composition of the surfactants used, the charge of the clay and its CEC value, and

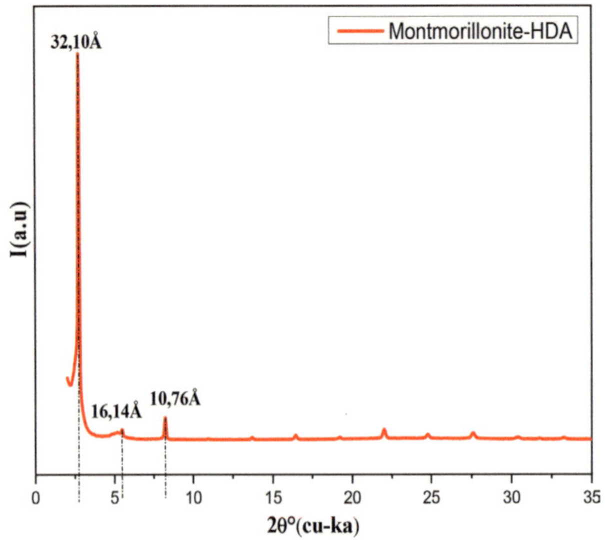

Figure 10.
Experimental XRD profile of montmorillonite-HDA specimen.

several arrangements and conformations of the molecules within the clay can be obtained.

Among the possible configurations of the molecule in the interlayer space, one corresponds to a conformation where the plane of the arrangement of the carbon atoms of the surfactant chain is parallel to the surface of the silicate layer and the other in a perpendicular position.

The thickness of the alkyl chain varies from 4.1 Å to 4.5 Å, while that of the cationic head NH^{3+} varies between 3.17 Å and 3.3 Å (**Figure 11**).

5.2.3 Radionuclide immobilization

Montmorillonite plays a critical role in radioactive waste management:

- Cesium (Cs): Distribution coefficients (Kd) up to 10^5 mL/g [46].
- Strontium (Sr): Adsorption capacities up to 180 mg/g [36, 56].
- Uranium (U): Removal efficiencies greater than 95% [47].

5.3 Successful applications in groundwater remediation

5.3.1 Permeable reactive barriers (PRBs)

Montmorillonite-based PRBs have been effectively employed for in situ groundwater remediation. For instance, a montmorillonite-zeolite PRB in Aznalcóllar, Spain, successfully reduced heavy metal concentrations (As, Zn, Cu, and Cd) in acid mine drainage-contaminated groundwater by over 99% over a five-year period [51].

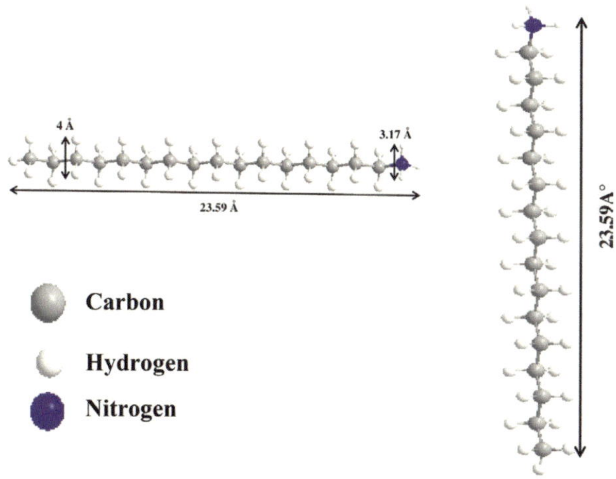

Figure 11.
Schematic representation of possible molecular conformations in the interlayer space.

5.3.2 Engineered geosorbents

Modified montmorillonites have shown enhanced performance in field applications. An example includes an organo-montmorillonite geosorbent that reduced BTEX (Benzene, Toluene, Ethylbenzene, Xylene) concentrations by more than 95% in a gasoline-contaminated aquifer in Michigan, USA [57].

5.3.3 Nanocomposite materials

Montmorillonite-based nanocomposites have demonstrated promising results in pilot-scale studies. A montmorillonite-iron oxide nanocomposite, for example, achieved over 99% removal of arsenic from groundwater in a field trial in West Bengal, India, maintaining its efficacy over a six-month period [58].

5.4 Challenges and future directions

Despite its outstanding properties, the application of montmorillonite in groundwater remediation faces several challenges:

- Long-term stability: The potential for colloidal transport of montmorillonite particles requires further investigation [47].

- Competitive adsorption: The presence of multiple contaminants and naturally occurring ions can affect removal efficiencies [54].

Future research should focus on:

- Developing novel modification techniques to enhance montmorillonite's selectivity and capacity for specific contaminants.

- Investigating the long-term performance and stability of montmorillonite-based remediation systems under various geochemical conditions.
- Exploring synergistic combinations of montmorillonite with other materials (e.g., biochar, nanomaterials) to improve remediation efficacy.
- Advancing modeling approaches to better predict montmorillonite's performance in complex, multi-contaminant groundwater systems.

In conclusion, montmorillonite's remarkable properties establish it as a highly versatile and effective material for groundwater decontamination. Its proven success in various remediation applications, coupled with ongoing research and development, underscores its potential as a critical tool in addressing global groundwater contamination challenges.

6. Geological barriers

Clay minerals play a vital role in the natural attenuation of contaminants within groundwater systems, functioning as geological barriers that significantly impede the migration of pollutants. These minerals, particularly through adsorption processes, can considerably reduce the spread of pollution without the need for human intervention. However, shifts in environmental conditions can compromise the stability of these natural barriers, potentially resulting in the release of previously immobilized contaminants [27].

6.1 Natural attenuation through clay mineral adsorption

The effectiveness of clay minerals in natural attenuation stems from their high adsorption capacity, which enables them to act as barriers against contaminant migration. This process, termed natural attenuation, involves various mechanisms by which clay minerals retard the movement of contaminants. These mechanisms include ion exchange, where cationic contaminants are replaced by naturally occurring cations in the clay's interlayers; surface complexation, which involves the formation of chemical bonds between contaminants and clay surfaces; physical adsorption through Van der Waals forces; and interlayer intercalation, where some contaminants enter the expandable spaces between clay layers. For example, a study conducted in the Chalk aquifer of southern England showed that a thin layer of clay-rich deposits, primarily composed of smectite and illite, effectively slowed the migration of a trichloroethene (TCE) plume for over 20 years. This clay layer reduced TCE concentrations from 1000 μg/L to less than 1 μg/L over a vertical distance of less than 1 meter [59].

6.1.1 Mechanism of contaminant retardation

Clay minerals retard contaminant migration through several mechanisms:

- Ion exchange: Cationic contaminants are exchanged with naturally occurring cations in clay interlayers.
- Surface complexation: Chemical bonds form between contaminants and clay surfaces.

- Physical adsorption: Van der Waals forces attract contaminants to clay surfaces.
- Interlayer intercalation: Some contaminants can enter the expandable interlayer spaces of certain clay minerals.

6.1.2 Quantifying retardation

The effectiveness of clay minerals in retarding the migration of contaminants can be quantitatively assessed using the retardation factor (R). This factor provides a measure of how much the movement of a contaminant is slowed down relative to the groundwater flow. The retardation factor is calculated using the formula (Eq. (5)):

$$R = 1 + \frac{\rho_b \cdot K_d}{\theta} \tag{5}$$

Where:

- (ρ_b) is the bulk density of the aquifer material (g/cm^3),
- (Kd) is the distribution coefficient, which reflects the affinity of the contaminant for the solid phase versus the aqueous phase (L/kg),
- (θ) is the porosity of the aquifer material (dimensionless, often expressed as a fraction).

6.1.2.1 Example of retardation factor calculation

To illustrate how the retardation factor can be applied, consider a case where cesium (Cs) is transported through bentonite, which is primarily composed of montmorillonite. The distribution coefficient (Kd) for cesium in bentonite is typically high due to strong adsorption. Assume the following values for a given aquifer:

- Bulk density (ρ_b) = 1.5 g/cm^3
- Distribution coefficient Kd = 300 L/kg
- Porosity θ = 0.3 (or 30%)

Plugging these values into the formula (Eq. (6)):

$$R = 1 + \frac{1.5\text{g}/cm^3 \cdot 300\text{L}/kg}{0.3} = 1501 \tag{6}$$

This result indicates that cesium migration through this bentonite is retarded by a factor of 1501 compared to the flow of groundwater. This means that cesium moves through the material approximately 1500 times slower than the groundwater.

6.1.2.2 Case studies and literature evidence

Several studies have demonstrated the application of the retardation factor in evaluating contaminant migration. For instance, Alshameri et al. [46] investigated cesium transport through bentonite, predominantly montmorillonite, and found retardation factors ranging from 2000 to 5000. This wide range indicates that cesium migration was thousands of times slower than groundwater flow, highlighting the significant retardation capacity of montmorillonite-rich clays in natural settings.

Similarly, research on the transport of other contaminants in various clay minerals has shown comparable effects. For example, in a study by Mejri et al. [31], the retardation factor for lead (Pb) in kaolinite was found to be around 1000, demonstrating its effectiveness in reducing the mobility of heavy metals in contaminated groundwater systems.

6.1.2.3 Implications for groundwater management

Understanding and quantifying the retardation factor is crucial for groundwater management, especially in the context of designing effective remediation strategies and predicting the movement of contaminants. High retardation factors suggest that clay minerals can serve as effective barriers in groundwater systems, reducing the risk of widespread contamination.

The retardation factor provides a valuable quantitative measure of how effectively clay minerals can slow down contaminant migration. The substantial retardation factors observed in studies involving bentonite and other clay minerals underscore their importance in groundwater protection and remediation efforts. Future research should continue to refine these measurements and explore the variability of retardation factors across different clay types and contaminant scenarios.

6.2 Stressed materials: Impact of environmental changes

While clay minerals can effectively immobilize contaminants under stable conditions, changes in the geochemical environment can lead to the release of previously adsorbed pollutants. This concept is known as "stressed materials."

6.2.1 Factors affecting stability of adsorbed contaminants

Several environmental factors can stress clay minerals and affect their contaminant retention capacity:

- *pH changes*: Alterations in pH can affect surface charge and contaminant speciation.
- *Redox conditions*: Changes in oxidation–reduction potential can alter contaminant mobility.
- *Ionic strength*: Variations in groundwater salinity can influence clay mineral structure and adsorption properties.
- *Temperature*: Thermal changes can affect adsorption equilibria and clay mineral structure.

- *Organic matter*: Introduction of organic compounds can compete for adsorption sites or form complexes with contaminants.

For example, a field study in the Aquia aquifer (Maryland, USA) revealed that the intrusion of high-ionic-strength water led to the release of naturally occurring arsenic from clay minerals. Arsenic concentrations in groundwater increased from less than 1 μg/L to over 100 μg/L due to ion exchange processes triggered by increased salinity [60].

6.2.2 Case studies of contaminant release

Several studies [60–65] have documented the release of contaminants from clay minerals under stressed conditions:

- *Acid mine drainage*: In the Zelazny Most tailings pond in Poland, the oxidation of sulfide minerals led to a decrease in pH from 8 to 2.3. This pH drop caused the release of previously adsorbed heavy metals (Cu, Zn, and Pb) from clay minerals, increasing their concentrations in groundwater by up to two orders of magnitude.

- *Nuclear waste repositories*: Experiments simulating conditions in nuclear waste repositories showed that an increase in temperature from 25–80°C led to a 30% reduction in cesium sorption on bentonite due to changes in clay mineral structure and sorption mechanisms [61].

- *Coastal aquifers*: In a study of a coastal aquifer in Barcelona (Spain), seawater intrusion increased the ionic strength of groundwater, leading to the release of adsorbed arsenic from clay minerals. Arsenic concentrations increased from less than 10 μg/L to over 50 μg/L in affected areas.

6.3 Managing stressed materials in groundwater systems

Understanding the potential for contaminant release under stressed conditions is crucial for effective groundwater management and remediation strategies.

6.3.1 Monitoring and prediction

Regular monitoring of key geochemical parameters, such as pH, redox potential, and major ion concentrations, is essential for identifying potential stressors. Geochemical modeling tools like PHREEQC can predict contaminant behavior under changing conditions. For instance, in the Netherlands, a comprehensive monitoring program was implemented in areas prone to seawater intrusion. By tracking chloride concentrations and redox conditions, authorities were able to predict and mitigate arsenic mobilization from clay minerals [62].

6.3.2 Remediation strategies for stressed systems

When dealing with stressed clay mineral systems, specialized remediation approaches may be necessary. These strategies include pH control through buffering agents, redox manipulation using electron donors or acceptors, and the installation of permeable reactive barriers to re-capture mobilized contaminants. For example, at a former mining site in Cornwall (UK), a permeable reactive barrier containing limestone was installed to address the mobilization of heavy metals caused by acid mine drainage. This barrier successfully increased the pH from 3.5 to 6.5 and reduced dissolved metal concentrations by over 99% [63].

6.4 Future research directions

To better understand and manage stressed clay mineral systems in groundwater, future research should focus on several key areas. These include long-term field studies to evaluate the stability of clay mineral barriers under various environmental stressors, the development of advanced monitoring techniques for early detection of clay mineral stress and contaminant mobilization, and the investigation of coupled processes that may influence contaminant behavior in stressed systems. Additionally, improving predictive models to account for complex interactions in multi-stressor scenarios and exploring novel remediation techniques specifically designed for stressed clay mineral systems will be crucial for effective groundwater protection and remediation efforts.

In conclusion, while clay minerals serve as essential natural barriers in groundwater systems, their effectiveness can be compromised under stressed conditions. Understanding the factors that lead to contaminant release and developing strategies to manage these stressed materials are critical for maintaining long-term groundwater protection and successful remediation.

7. Expanded conclusion

This expanded conclusion synthesizes the key findings of the chapter, addresses current challenges, and outlines future research directions in the field of clay mineral-based groundwater remediation.

7.1 Summary of key findings

7.1.1 Remarkable properties of clay minerals

Clay minerals, especially montmorillonite, exhibit exceptional capabilities for groundwater decontamination and protection. These minerals possess a high surface area—up to 800 m^2/g for montmorillonite—providing numerous adsorption sites for contaminants. Their significant cation exchange capacity, ranging from 80 to 150 meq/100 g for montmorillonite, facilitates effective removal of cationic contaminants. The swelling properties of clay minerals enhance their ability to intercept contaminants and create low-permeability barriers. Additionally, their structural flexibility enables the intercalation of various organic and inorganic pollutants.

7.1.2 Versatility in contaminant removal

Clay minerals have demonstrated efficacy in removing a broad spectrum of groundwater contaminants. They achieve removal efficiencies exceeding 90% for heavy metals such as Pb, Cd, Cu, and Zn [6]. Their adsorption capacities can reach up to 200 mg/g for organic pollutants, including pesticides and pharmaceuticals [32].

For radionuclides, distribution coefficients $((K_d))$ can be as high as $(10^5\ mL/g)$ for cesium, indicating that cesium migration is significantly retarded compared to groundwater flow [46].

7.1.3 Natural attenuation and engineered solutions

Clay minerals contribute to groundwater protection through natural geological barriers that retard contaminant migration. Engineered solutions such as permeable reactive barriers and geosorbents utilize these minerals to enhance contaminant removal. Nanocomposite materials, which integrate clay minerals, further improve removal efficiency and selectivity.

7.1.4 Current challenges and research needs

7.1.4.1 Complex geochemical interactions

There is a need for a deeper understanding of competitive adsorption in multi-contaminant systems, as well as the influence of natural organic matter on clay mineral performance. The long-term stability of clay minerals under various geochemical conditions also requires investigation. Future research should focus on developing comprehensive geochemical models that integrate clay mineral-contaminant interactions with site-specific conditions.

7.1.4.2 Upscaling and field implementation

Bridging the gap between laboratory studies and field-scale applications is a significant challenge. Addressing heterogeneity in subsurface environments and optimizing deployment methods for in situ remediation are crucial. More long-term field trials are needed to validate the performance and longevity of clay mineral-based remediation systems.

7.1.4.3 Emerging contaminants

Assessing the efficacy of clay minerals against emerging pollutants, such as PFAS and microplastics, is essential. Developing modification techniques to enhance selectivity for specific contaminants is also necessary. Research should investigate the interaction mechanisms between clay minerals and emerging contaminants to design targeted remediation strategies.

7.2 Future research directions

7.2.1 Advanced material design

Future research should focus on developing "smart" clay-based materials with stimuli-responsive properties. Exploring synergistic combinations of clay minerals with other materials, such as biochar or zero-valent iron, could enhance remediation performance. Computational methods may also be used for the rational design of clay-based adsorbents. For example, recent work on pH-responsive organoclays has shown potential in controlled release applications, which could be adapted for groundwater remediation [66].

7.2.2 In situ monitoring and real-time optimization

Developing sensors for real-time monitoring of clay mineral performance in groundwater systems is crucial. Implementing machine learning algorithms for predictive modeling and optimization of remediation strategies could enhance effectiveness. A pilot study using fiber optic sensors embedded in clay-based permeable reactive barriers has demonstrated real-time monitoring of contaminant breakthrough, enabling timely barrier regeneration [67].

7.2.3 Circular economy approaches

Research should explore methods for regenerating and reusing spent clay minerals, as well as beneficial reuse options for contaminated clay materials. Assessing the life cycle impacts of clay mineral-based remediation technologies is also important. Recent research has shown that heat treatment of metal-loaded montmorillonite can recover valuable metals while regenerating the clay for reuse in water treatment [68].

7.2.4 Integration with biological remediation

Investigating clay mineral-microbe interactions for enhanced bioremediation is an area of interest. Developing clay-based carriers for beneficial microorganisms in groundwater systems could improve outcomes. A study combining montmorillonite with sulfate-reducing bacteria demonstrated synergistic effects in removing heavy metals from acid mine drainage, with the clay providing both a support matrix for the bacteria and additional metal adsorption capacity [69].

7.2.5 Holistic groundwater management approaches

7.2.5.1 Multi-barrier concepts

Integrating clay mineral-based technologies with other remediation techniques to create robust, multi-barrier systems for groundwater protection is essential. For instance, a comprehensive remediation strategy for a contaminated industrial site in Germany combined a montmorillonite-based permeable reactive barrier with phytoremediation and enhanced natural attenuation, resulting in over a 99% reduction in contaminant concentrations over a five-year period [70].

7.2.5.2 Adaptive management strategies

Developing flexible, adaptive management approaches that can respond to changing environmental conditions and emerging contamination challenges is critical. The implementation of a "treatment train" approach at a former military site in the USA, incorporating clay-based sorption, advanced oxidation, and bioremediation, allowed for dynamic adjustment of treatment processes based on monitoring data, leading to cost-effective and sustainable long-term remediation [71].

7.2.5.3 Policy and regulatory frameworks

Advocating for policies and regulations that promote the use of sustainable, clay mineral-based remediation technologies and support long-term monitoring and management of remediated sites is necessary for advancing the field.

7.3 Concluding remarks

Clay minerals, particularly montmorillonite, emerge as valuable assets in the battle against groundwater contamination. Their remarkable properties and versatility in contaminant removal offer promising solutions for both natural attenuation and engineered remediation strategies. However, realizing their full potential requires addressing current challenges through interdisciplinary research and holistic management approaches. By integrating advanced material design, in situ monitoring technologies, and sustainable management practices, clay mineral-based remediation can play a crucial role in safeguarding groundwater resources for future generations. Continued exploration of clay mineral-contaminant interactions will unlock new possibilities for utilizing these ubiquitous natural materials in our ongoing efforts to protect and restore the world's precious groundwater resources.

Acknowledgements

The findings presented in this study are the product of an extensive research endeavor, made possible through collaborative efforts and rigorous data analysis. The authors acknowledge the use of CLAUDE AI for the language polishing of the manuscript.

Conflict of interest

The authors declare no conflict of interest.

Notes/thanks/other declarations

The research presented in this chapter would not have been possible without the tireless efforts of the many undergraduate and graduate students who have worked in Dr. Walid Oueslati's research group over the years.

Author details

Chadha Mejri[1] and Walid Oueslati[1,2]*

1 Faculty of Sciences of Bizerte, Physics Department, University of Carthage, Bizerte, Tunisia

2 Faculty of Sciences of Bizerte, Resources, Materials and Ecosystem (RME), University of Carthage, Bizerte, Tunisia

*Address all correspondence to: walidoueslati@ymail.com

References

[1] Foster S, Chilton PJ. Groundwater: The processes and global significance of aquifer degradation. Philosophical Transactions of the Royal Society of London. Series B: Biological Sciences. 2003;**358**(1440):1957-1972

[2] Famiglietti JS. The global groundwater crisis. Nature Climate Change. 2014;**4**(11):945-948

[3] Bergaya F, Lagaly G. editors. Handbook of Clay Science. 2nd ed. Amsterdam, Netherlands: Elsevier; 2013

[4] Oueslati W, Meftah M. Discretization of the water uptake process of Na-montmorillonite undergoing atmospheric stress: XRD Modeling approach. Advances in Materials Science and Engineering. 2018;**2018**(1):5219624

[5] Oueslati W et al. Impact of uniaxial mechanical perturbation on structural properties and Smectite porosity features: Ion exchanger efficiency and adsorption performance fate. Advances in Civil Engineering. 2022;**2022**:1-16. DOI: 10.1155/2022/4441705

[6] Bhattacharyya KG, Gupta SS. Adsorption of a few heavy metals on natural and modified kaolinite and montmorillonite: A review. Advances in Colloid and Interface Science. 2008;**140**(2):114-131

[7] Oueslati W, Rhaiem HB, Amara ABH. XRD investigations of hydrated homoionic montmorillonite saturated by several heavy metal cations. Desalination. 2011;**271**(1-3):139-149

[8] Ammar M, Oueslati W, Ben Rhaiem H, Amara BH. Quantitative XRD analysis of the dehydration–hydration performance of (Na+, Cs+) exchanged smectite. Desalination and Water Treatment. 2014;**52**(22-24):4314-4333

[9] Uddin MK. A review on the adsorption of heavy metals by clay minerals, with special focus on the past decade. Chemical Engineering Journal. 2017;**308**:438-462

[10] Liu Y et al. Adsorption of levofloxacin onto an iron-pillared montmorillonite (clay mineral): Kinetics, equilibrium and mechanism. Applied Clay Science. 2019;**174**:77-84

[11] Tournassat C, Bourg IC, Steefel CI, Bergaya F. Surface properties of clay minerals. In: Developments in Clay Science. Vol. 6. Amsterdam, Netherlands: Elsevier; 2015. pp. 5-31

[12] Sposito G, Skipper NT, Sutton R, Park SH, Soper AK, Greathouse JA. Surface geochemistry of the clay minerals. Proceedings of the National Academy of Sciences. 1999;**96**(7):3358-3364

[13] Tombacz E, Szekeres M. Colloidal behavior of aqueous montmorillonite suspensions: The specific role of pH in the presence of indifferent electrolytes. Applied Clay Science. 2004;**27**(1-2):75-94

[14] Bourg IC, Sposito G, Bourg AC. Modeling the acid–base surface chemistry of montmorillonite. Journal of Colloid and Interface Science. 2007;**312**(2):297-310

[15] Bradl HB. Adsorption of heavy metal ions on soils and soils constituents. Journal of Colloid and Interface Science. 2004;**277**(1):1-18

[16] Oueslati W, Ben Rhaiem H, Amara ABH. Selectivity of

Na-montmorillonite versus concentration of two competitive bivalent cations (Cu^{2+}, Pb^{2+}): Quantitative XRD investigation. Advances in Materials Science and Engineering. 2009;**2009**:385673. DOI: 10.1155/2009/385673

[17] Grim RE. Clay mineralogy McGraw Hill Book Company. Journal of Geological Society of India. 2005;**66**:179-171

[18] Cases JM, Berend I, Besson G, Francois M, Uriot JP, Thomas F, et al. Mechanism of adsorption and desorption of water vapor by homoionic montmorillonite. 1. The sodium-exchanged form. Langmuir. 1992;**8**(11):2730-2739

[19] Oueslati W, Ben Rhaiem H, Karmous MS, Naaman S, Amara BH, A. Study of the structural evolution and selectivity of Wyoming montmorillonite in relation with the concentration of Cu2+ and Ni2+. Zeitschrift für Kristallographie Supplement. 2006;**2006**:425-429

[20] Oueslati W, Karmous MS, Rhaiem HB, Lanson B, Amara ABH. Effect of interlayer cation and relative humidity on the hydration properties of a dioctahedral smectite. Zeitschrift fur Kristallographie, Supplement. 2007;**2**(26):417-422

[21] Lee SM, Tiwari D. Organo and inorgano-organo-modified clays in the remediation of aqueous solutions: An overview. Applied Clay Science. 2012;**59**:84-102

[22] Liang L, Li X, Lin Z, Tian C, Guo Y. The removal of Cd by sulfidated nanoscale zero-valent iron: The structural, chemical bonding evolution and the reaction kinetics. Chemical Engineering Journal. 2020;**382**:122933

[23] Dou X et al. Enhanced removal of U(VI) by montmorillonite modified with polydopamine and zero-valent iron nanoparticles. Journal of Environmental Chemical Engineering. 2021;**9**(1):104955

[24] Caraballo MA, Serna A, Macías F, Pérez-López R, Ruiz-Cánovas C, Richter P, et al. Uncertainty in the measurement of toxic metals mobility in mining/mineral wastes by standardized BCR® SEP. Journal of Hazardous Materials. 2018;**360**:587-593

[25] Wang D, Wang H, Larsson S, Benzerzour M, Maherzi W, Amar M. Effect of basalt fiber inclusion on the mechanical properties and microstructure of cement-solidified kaolinite. Construction and Building Materials. 2020;**241**:118085

[26] Mejri C, Oueslati W, Amara ABH. From coupling second-order stresses to understanding and predicting the structural response of a Dioctahedral Smectite. Clays and Clay Minerals. 2023;**71**(5):513-538

[27] Mejri C, Oueslati W, Amara ABH. Structure and reactivity assessment of dioctahedral montmorillonite during provoked variable sequential cation exchange process via XRD modelling approach. Applied Surface Science Advances. 2023;**15**:100403

[28] Mollah MYA et al. A review of arsenic removal technologies for contaminated groundwaters. Environmental International. 2000;**26**(3):291-295

[29] Oueslati W, Rhaiem HB, Amara ABH. Effect of relative humidity constraint on the metal exchanged montmorillonite performance: An XRD profile modeling approach. Applied Surface Science. 2012;**261**:396-404

[30] Mejri C, Oueslati W, Amara ABH. How the solid/liquid ratio affects the cation exchange process and porosity in the case of dioctahedral smectite: Structural analysis? Adsorption Science and Technology. 2021;**2021**:9732092

[31] Mejri C, Oueslati W, Amara ABH. Structural alteration, hydration stability, heavy metal removal efficiency, and montmorillonite porosity fate by coupling the soil solution ph and a thermal gradient. Adsorption Science and Technology. 2022;**2022**:4421932

[32] Jiang MQ et al. Adsorption of Pb(II), Cd(II), Ni(II) and Cu(II) onto natural kaolinite clay. Desalination. 2010;**252**(1-3):33-39

[33] Krestou A, Xenidis A, Panias D. Mechanism of aqueous uranium (VI) uptake by natural zeolitic tuff. Minerals. 2018;**8**(10):449

[34] Celis R et al. Sorption of atrazine and simazine by model associations of soil colloids. Soil Science Society of America Journal. 2000;**64**(5):1656-1661

[35] Liu C et al. Retardation of lead migration in kaolinite: Implications for groundwater contamination. Environmental Science and Technology. 2015;**49**(4):2267-2275

[36] Zhu R et al. Adsorbents based on montmorillonite for contaminant removal from water: A review. Applied Clay Science. 2016;**123**:239-258

[37] Oueslati W. Effect of soil solution pH during the tetracycline intercalation on the structural properties of a dioctahedral smectite: Microstructural analysis. Journal of Nanomaterials. 2019;**2019**(1):7414039

[38] Peng H et al. Contribution of hydrophobic effect to the sorption of tetracycline and ciprofloxacin on graphene oxides. Journal of Hazardous Materials. 2020;**384**:121412

[39] De Oliveira T et al. Adsorption of diclofenac onto organoclays: Effects of surfactant and environmental (pH and temperature) conditions. Journal of Hazardous Materials. 2017;**323**:558-566

[40] Aghadadashi V, Neyestani MR, Mehdinia A, Bakhtiari AR, Molaei S, Farhangi M, et al. Spatial distribution and vertical profile of heavy metals in marine sediments around Iran's special economic energy zone; arsenic as an enriched contaminant. Marine Pollution Bulletin. 2019;**138**:437-450

[41] Alshameri A, He H, Zhu J, Xi Y, Zhu R, Ma L, et al. Adsorption of ammonium by different natural clay minerals: Characterization, kinetics and adsorption isotherms. Applied Clay Science. 2018;**159**:83-93

[42] Yu AJ, Lin X, Zhu J, He H, Li L. Environmental effects on ammonium adsorption onto clay minerals: Experimental constraints and applications. Applied Clay Science. 2023;**246**:107165

[43] Eturki S, Ayari F, Jedidi N, et al. Use of clay mineral to reduce ammonium from wastewater. Effect of various parameters. Surface Engineering and Applied Electrochemistry. 2012;**48**:276-283. DOI: 10.3103/S1068375512030064

[44] Kozak M, Domka L. Adsorption of the quaternary ammonium salts on montmorillonite. Journal of Physics and Chemistry of Solids. 2004;**65**(2-3):441-445

[45] Jing QX, Chai LY, Huang XD, Tang CJ, Huan GUO, Wei WANG.

Behavior of ammonium adsorption by clay mineral halloysite. Transactions of Nonferrous Metals Society of China. 2017;**27**(7):1627-1635

[46] Alshameri A, Wei X, Wang H, Fuguo Y, Chen X, He H, et al. A review of the role of natural clay minerals as effective adsorbents and an alternative source of minerals. In: Minerals (Edited Book), Chapter 3. London, UK: IntechOpen; 2019. pp. 49-64. DOI: 10.5772/intechopen.87260

[47] Steefel CI et al. Cesium migration in Hanford sediment: A multisite cation exchange model based on laboratory transport experiments. Journal of Contaminant Hydrology. 2003;**67**(1-4):219-246

[48] Xie S, Huang L, Su C, Yan J, Chen Z, Li M, et al. Application of clay minerals as adsorbents for removing heavy metals from the environment. Green and Smart Mining Engineering. 2024;1(3):249-261

[49] Bradbury MH, Baeyens B. Experimental measurements and modeling of sorption competition on montmorillonite. Geochimica et Cosmochimica Acta. 2005;**69**(17):4187-4197

[50] Galamboš M et al. Adsorption equilibrium and kinetic studies of strontium on Mg-bentonite, Fe-bentonite and illite/smectite. Journal of Radioanalytical and Nuclear Chemistry. 2019;**318**(3):2213-2224

[51] Zavarin M et al. Np(V) and Pu(V) ion exchange and surface-mediated reduction mechanisms on montmorillonite. Environmental Science and Technology. 2012;**46**(5):2692-2698

[52] Pusch R, Kasbohm J, Knutsson S, Yang T, Nguyen-Thanh L. The role of smectite clay barriers for isolating high-level radioactive waste (HLW) in shallow and deep repositories. Procedia Earth and Planetary Science. 2015;**15**:680-687

[53] Bourg IC, Tournassat C. Self-diffusion of water and ions in clay barriers. In: Developments in Clay Science. Vol. 6. Amsterdam, Netherlands: Elsevier; 2015. pp. 189-226

[54] Brigatti MF, Galan E, Theng BKG. Structure and mineralogy of clay minerals. In: Developments in Clay Science. Vol. 5. Amsterdam, Netherlands: Elsevier; 2013. pp. 21-81

[55] Liu Y, He LQ, Jiang YJ, Sun MM, Chen EJ, Lee FH. Effect of in situ water content variation on the spatial variation of strength of deep cement-mixed clay. Géotechnique. 2019;**69**(5):391-405

[56] Forgionny A, Acelas NY, Ocampo-Pérez R, Padilla-Ortega E, Leyva-Ramos R, Flórez E. Understanding mechanisms in the adsorption of lead and copper ions on chili seed waste in single and multicomponent systems: A combined experimental and computational study. Environmental Science and Pollution Research. 2021;**28**:23204-23219

[57] Chen J, Montañez IP, Qi Y, Shen S, Wang X. Strontium and carbon isotopic evidence for decoupling of pCO2 from continental weathering at the apex of the late Paleozoic glaciation. Geology. 2018;**46**(5):395-398

[58] Mayacela-Rojas CM, Molinari A, Cortina JL, Gibert O, Ayora C, Tavolaro A, et al. Removal of transition metals from contaminated aquifers by PRB technology: Performance comparison among reactive materials. International Journal of Environmental Research and Public Health. 2021;**18**(11):6075

[59] Zhao B, Huang F, Zhang C, Huang G, Xue Q, Liu F. Pollution characteristics of aromatic hydrocarbons in the groundwater of China. Journal of Contaminant Hydrology. 2020;**233**:103676

[60] Mukhopadhyay R, Bhaduri D, Sarkar B, Rusmin R, Hou D, Khanam R, et al. Clay–polymer nanocomposites: Progress and challenges for use in sustainable water treatment. Journal of Hazardous Materials. 2020;**383**:121125

[61] Hiscock KM, Grischek T. Attenuation of groundwater pollution by bank filtration. Journal of Hydrology. 2002;**266**(3-4):139-144

[62] Welch AH, Westjohn DB, Helsel DR, Wanty RB. Arsenic in ground water of the United States: Occurrence and geochemistry. Groundwater. 2000;**38**(4):589-604

[63] Boudreau V, Chen R, Edwards A, Sulaimain M, Maddox PS. PP2A-B55/SUR-6 collaborates with the nuclear lamina for centrosome separation during mitotic entry. Molecular Biology of the Cell. 2019;**30**(7):876-886

[64] Wallis I, Prommer H, Berg M, Siade AJ, Sun J, Kipfer R. The river–groundwater interface as a hotspot for arsenic release. Nature Geoscience. 2020;**13**(4):288-295

[65] Tsunashima A, Itoh H, Katano T. Effects of temperature and phytoplankton community composition on subitaneous and resting egg production rates of Acartia omorii in Tokyo Bay. Scientific Reports. 2021;**11**(1):7959

[66] Zhang JR, Xu MD, Christidis GE, Zhou CH. Clay minerals in drilling fluids: Functions and challenges. Clay Minerals. 2020;**55**(1):1-11

[67] Li C, Zhu N, Yang S, He X, Zheng S, Sun Z, et al. A review of clay based photocatalysts: Role of phyllosilicate mineral in interfacial assembly, microstructure control and performance regulation. Chemosphere. 2021;**273**:129723

[68] Wang L, Cho DW, Tsang DC, Cao X, Hou D, Shen Z, et al. Green remediation of As and Pb contaminated soil using cement-free clay-based stabilization/solidification. Environment International. 2019;**126**:336-345

[69] Karna RR, Hettiarachchi GM, Van Nostrand J, Yuan T, Rice CW, Assefa Y, et al. Microbial population dynamics and the role of sulfate reducing bacteria genes in stabilizing Pb, Zn, and Cd in the terrestrial subsurface. Soil Systems. 2018;**2**(4):60

[70] Ma B, Charlet L, Fernandez-Martinez A, Kang M, Madé B. A review of the retention mechanisms of redox-sensitive radionuclides in multi-barrier systems. Applied Geochemistry. 2019;**100**:414-431

[71] Khan S, Ajmal S, Hussain T, Rahman MU. Clay-based materials for enhanced water treatment: Adsorption mechanisms, challenges, and future directions. Journal of Umm Al-Qura University for Applied Sciences. 2023;**2023**:1-16

Chapter 2

Assessment of Groundwater Variation and Identifying Influencing Factors over Abbay River Basin, East Africa

Agegnehu Kitanbo Yoshe

Abstract

The effect of climate change and anthropogenic activities on groundwater storage has gained attention recently across the globe. The increasing water demand and overexploitation of surface water add extra pressure on groundwater. This paper estimated groundwater storage with spatiotemporal variation by using the Gravity Recovery and Climate Experiment satellite mission with the Global Assimilation System. The investigation revealed that there is a significant variation in groundwater storage following seasonal variation during the study period. We found that GRACE and GLDAS datasets can be combined effectively to evaluate the long-term GWS in large-scale basins with limited hydrological datasets in an area.

Keywords: RFM, groundwater storage, R-studio, climatic parameter, GRACE, GLDAS

1. Introduction

All kinds of water are vital to life, but groundwater is very important due to its unique nature. It was expected to be the purest water among all other water resources because it was filtered and cleaned by soil layers from many contaminants. Subsequently, individuals are utilizing groundwater to fulfill their needs [1, 2]. Users of groundwater include domestic use, agricultural use, industrial use, recreational, hydroelectric, and environmental activities. The quantity and quality of water resources are key concerns. Mismanagement, misuse, pollution, and scarcity of this life-sustaining resource pose a serious problem [3]. It is a growing threat to water supply, agriculture, food security, industrial development, and the ecosystems on which they depend [1, 4]. The number of serious issues and completions among the different water users has increased as a result of growing populations and different activities related to water. Due to the fact that water is primarily a common resource worldwide. Water is also a common resource worldwide and a number of serious issues among different water users has increased as a result of growing populations and climate change.

Water located in the saturated zone was groundwater [5], and the availability of groundwater was affected by different geo-environmental factors [6]. Groundwater uses geological strata as reservoirs for storage, while the transmissivity and storage affect the capacity of the exploitation of these natural resources in the given geological location. More than 2 billion people use groundwater as a major source of water supply [5, 7]. Protection and management of groundwater were mostly done by identifying the potential zones of groundwater level [8]. This mostly estimates groundwater resources by using indicators, and it is difficult to fully explain the groundwater potential in an area.

The hydrological water balance is the combination of surface and groundwater [9]. In particular, groundwater balance assessment was the most challenging because of data scarcity [10]. On a regional and continental scale, routine monitoring of groundwater with a network of boreholes is laborious, costly, and takes time. The irregularity of temporal data, lake of monitoring well, and the complexity of sub-soil and recharge methods make groundwater storage more complex. Additionally, law enforcement agencies occasionally impose restrictions on the hydrological data sources. To overcome this problem, different scholars try to solve the problem by using direct drilling bore wells, traditional methods [6], indirect methods such as ArcGIS-based and remote sensing methods of groundwater potential assessment were used worldwide [11, 12], frequency ratio methods [13], multi-criteria decision-making analysis [14], evidence weight methods [15], regression logistic methods [6], evidence belief function methods [16], and wetspass method [17]. However, it is very difficult to install observation well in mountainous and arid regions, which leads to serious problems with GWS information and inaccurate information from indirect methods. Due to the geographical and temporal constraints that impede comprehensive and precise groundwater measurements over wide geographic areas, it is challenging to estimate changes in groundwater storage [18]. But in order to assess water storage under all possible terrestrial scenarios, GRACE sought to compare temporal and spatial variations in the Earth's gravitational field. In contrast to other innovations, such as radiometers and radar, GRACE is able to detect water variations and groundwater with greater accuracy than 1 cm of similar water heights without much effort [19]. In contrast to other innovations, such as radiometers and radar, GRACE is able to detect water variations and groundwater with greater accuracy than 1 cm of similar water heights without much effort [19]. Groundwater reservoir changes can now be evaluated with the help of GRACE [19]. The water storage variation in an area from GRACE data is an estimate of changes in bulk storage capacity, and the data cannot be differentiated. Therefore, GRACE data must be interpreted with the GLDAS hydrological model to evaluate groundwater storage.

The majority of groundwater assessments in Ethiopia have been carried out through field surveys, which can be laborious to manage in terms of time and resources or limited in scope when done locally [20]. In this study area, the distribution of water to different stakeholder groups in appropriate quantities and quality has grown more complicated as a result of the growing demands for water for residential and commercial use, agriculture, hydropower generation, the environment, and other uses. Excessive groundwater extraction, especially for agricultural irrigation, has led to a dramatic decline in groundwater storage. Groundwater has also declined due to the extremely slow recharge mechanism [21, 22]. Traditional methods for measuring groundwater storage and fluctuations in the study area are tedious and time-consuming. Also, monitoring key wells in such a large area is costly and time-consuming. A number of studies have shown extensive applications of Gravity Recovery and Climate Experiment (GRACE), and TerraClimate data for the analysis of groundwater anomalies [23]. The GRACE data is useful for scientists and researchers to use in the field of hydrology and water resources [24]. From the review of past studies, it is clear that no attempts have been made to study the spatiotemporal variability of groundwater in the Abbay river basin using GRACE and Land Data Assimilation System (GLDAS) datasets. The general objective of this study is to estimate changes in groundwater storage by estimating soil moisture change, surface water change, and canopy water storage from GLDAS, followed by removing them from GRACE terrestrial water storage. The impact of GRACE-based application as a tool for groundwater resource management in the Abbay river basin was evaluated. It shows the application of GRACE in groundwater monitoring techniques and strategies for the sustainability of groundwater resources. The study also evaluates the spatial and temporal distribution of groundwater storage based on terrestrial water storage, canopy water storage, soil moisture storage, and surface runoff. Additionally, the monthly and annual groundwater storage variation to support sustainable groundwater monitoring and the impact of climatic change on groundwater storage for the study area were addressed. This study also describes the impact of remotely sensed groundwater storage variation and observance practices on decision-making and conventional modeling practices.

2. Methods and material

2.1 Description of the study area

The study area is located in the northwestern part of Ethiopia at $7^{O}40'$N and $12^{O}51'$N latitude and $34^{O}25'$E and $39^{O}49'$E longitude, with an area of approximately 176,200 km^2 and an elevation difference of 490–4266 m ams. The river starts in the high mountainous part of Ethiopia from Lake Tana and is used as a contributor to the Nile River. The upstream part of the river basin is dominated by mountainous landscapes, and most of the downstream areas are relatively flat or gently undulating. There were varying climatic zones in the river basin due to environmental conditions. The maximum temperature of the river basin ranges from 28 to 38°C, and the minimum temperature is 15–20°C downstream. Generally, rainfall in the study area ranges between 787 and 2200 mm per year, and the lowest rainfall recorded was less than 100 mm per year (**Figure 1**).

2.2 Data collection

Different datasets were used to estimate groundwater storage in the study area, such as terrestrial water storage (TWS) assessed from GRACE with spatial resolution 1Ox1O, soil moisture (SMS), surface runoff (Qs), canopy water storage (CWS), temperature (T), evapotranspiration (ET) from GLDAS processed at NOAH (0.25^{O}x0.25^{O}), and resamples to 1^{O}x1^{O} for further hydrological data analysis. Rainfall data from Terra-Climate are at 0.5^{O} resolution and resamples to 1^{O}x1^{O}. Data utilized for this study were collected from January 2010 to December 2020.

2.2.1 GRACE dataset

For tracking mass redistribution of the earth by monitoring changes in gravitational force, GRACE satellites were developed by NASA and the German Aerospace Center in March 2002 [25]. This dataset contains different errors, such as noise error

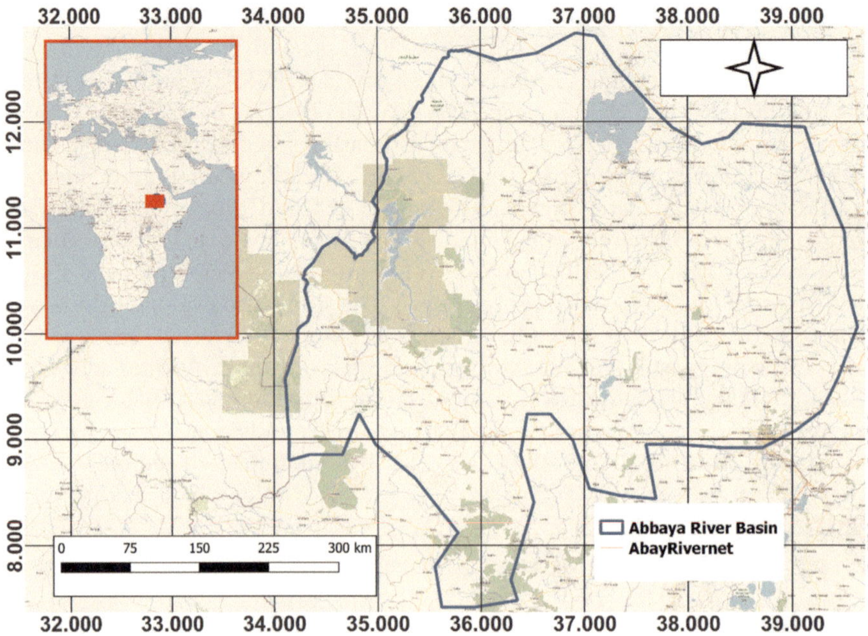

Figure 1.
Study area location map.

in the satellite for satellite microwave ranging measurements, accelerometer error, error in the ultra-stable oscillator, and orbit error in the form of stoke coefficient represented by and both for integer degree of and integer order of $0 \leq m \leq l$ [26]. To minimize measurement errors and noise, different filtering processes, such as a DE-striping filter, a wide Gaussian averaging filter, and a low-pass spectral filter, are utilized [27]. To improve the accuracy between the University of Texas Center for Space Research (CSR), the GeoForschungsZentrum (GFZ) Potsdam, and the Jet Propulsion Laboratory (JPL) for the estimation of TWS in the river basin, the cross-correlation coefficients between them were adopted to calculate the weights of the datasets [28].

2.2.2 GLDAS dataset

A Global Land Data Assimilation System (GLDAS) has been developed to ingest satellite and ground-based observational data by the National Oceanic and Atmospheric Administration and the National Center for Environmental Prediction, which simulates hydrologic components [29]. For the estimation of the change in groundwater storage from GRACE terrestrial water storage, the values of surface water storage (SWE), soil moisture storage (SMS), canopy water storage (CWS), and snow water storage (SWS) need to be deducted from the land surface model of NASA's Global Land Data Assimilation System (GLDAS version) at $1^{O}x1^{O}$ spatial resolution.

2.2.3 Terra-climate

Change in groundwater storage is influenced by the climatological conditions of an area, and knowing and understanding the natural characteristics of climatic datasets is important, can be conceptualized, and has an impact on runoff, infiltration, and groundwater recharge [30]. The monthly and climatic water balance for terrestrial water surfaces was developed from three global gridded climate datasets found at high spatiotemporal resolution and made publicly available through an unrestricted data repository of Idaho's Northwest Knowledge Network University. The climatological dataset for this study was gathered from the Terra-Climatic dataset [31]. The dataset was available from January 1, 1958 to December 1, 2021 at 0.5^{O} grid. The mean monthly precipitation and temperature from 2010 to 2020 were collected for this study and prepared for further data analysis.

2.3 Change in groundwater storage estimation for the study area

The inflow of water and water outflow from the system can be described by using the water balance equation. The water balance equation is difficult and complex because water balance components have uncertainties in measurement [32]. The change in terrestrial water storage data observed by GRACE includes vertical water storage components of the aquifer, which include soil moisture, surface runoff, canopy water storage, and groundwater storage, which are presented in eq. (1) [33].

$$\Delta TWS = \Delta GWS + \Delta SMS + \Delta SWE + \Delta CWS + \Delta Q_s \tag{1}$$

Where ΔTWS is the change in terrestrial water storage.
ΔGWS is the change in groundwater storage.
ΔSMS is the change in soil moisture storage.
ΔSWS is the change in snow water storage.
ΔCWS is the change in canopy water storage.
ΔQ_s is the change in surface water runoff.
By subtracting ΔSWS, ΔSMS, ΔCWS, ΔQ_s from observed GLDAS TWS yielding ΔGWS [33], as presented in eq. (2). The study area is semiarid, and with a warm climate, changes in snow water storage were ignored for this study.

$$\Delta GWS = \Delta TWS - (\Delta SMS + \Delta SWE + \Delta CWS + \Delta Q_s) \tag{2}$$

2.4 Statistical data analysis

Both parametric and nonparametric tests are used to examine the temporal variation. Nonparametric tests have an advantage over parametric tests in that they can be used for non-normal time series [34].

1. *Sen's slope estimator*: Sen's slant assessor determined the pattern's greatness and information pair was determined as:

$$Q_i = \frac{X_j - X_k}{j - k} \text{ where : (j > k) (for i = 1, 2, 3, ... N)} \tag{3}$$

Where X_k and X_j are the data values in years k and j, for j < k. If there are n values of X_i in a time series, one gets as many as $N = n * \frac{n-1}{2}$ to estimate the slope Q_i. Estimation of Sen's slope is the median of these N values of the slope S_i. The N values and Si were ranked from the smallest to the largest and Sen's carried out as follows.

The median of these the N value and Si is Sen's slope estimator, which is shown as follows:

$$Q_i = \begin{cases} Q_{\left[\left(\frac{N+1}{2}\right)\right]} \ if\ N\ is\ odd \\ \dfrac{Q_{\frac{N}{2}} + Q_{\frac{(N+2)}{2}}}{2} \ if\ N\ is\ even \end{cases} \tag{4}$$

When N is an even number and Sen's slope estimator is calculated as $Q_{med} = (Q_{(\frac{N}{2})} + \frac{Q_{\left(\frac{(N+2)}{2}\right)}}{2})$ if N appears odd and it was considered as $Q_{med} = Q_{\frac{N}{2}} + Q_{\frac{N+2}{2}}$ if N is even. Finally, Q_{med} is calculated by a two-side test at a 100% confidence interval, and then, a slope was calculated by a nonparametric test. The positive value of $Q_{i.}$ represents an increasing trend, and the negative value shows a decreasing trend in the time series.

2.5 Assessment of model performance methods

The downscaled terrestrial water storage results were computed using four distinct statistical metrics: R, NSE, MAE, and RME. The mathematical representation of MAE, NSE, R and RME is shown in eqs. 5–8 below. The values of RMSE and MAE close to 0 demonstrate the perfect model, and the value NSE and R closer to 1 shows the perfect model.

$$RMSE = \sqrt{\frac{\sum_{i=1}^{N}(X_i - Y_i)^2}{N}} \tag{5}$$

$$NSE = 1 - \frac{\sum_{i=1}^{N}(Y_i - \bar{Y})^2}{\sum_{i=1}^{N}(X_i - \bar{X})^2} \tag{6}$$

$$R = \frac{\sum_{i=1}^{N}(X_i - \bar{X})(Y_i - \bar{Y})}{\sqrt{\sum_{i=1}^{N}(X_i - \bar{X})^2}\sqrt{\sum_{i=1}^{N}(Y_i - \bar{Y})^2}} \tag{7}$$

$$MAE = \frac{1}{N}\left(\sum_{i=1}^{N}|Y_i - X_i|\right) \tag{8}$$

Where X_i and Y_i shows two independent datasets with the mean values of X and Y. The input terrestrial water storage is represented by X_i, and the predicted value of a random forest model is represented by Y_i, while N represents the total number of samples.

3. Result

3.1 Liquid water equivalent from GRACE dataset

The combination of the Duan de-striping method with a Gaussian filter with a radius of 150 km, 200 km, and 300 km was used to filter high-frequency noises. After using the Duan algorithm to remove the model stripe signal errors, our study area showed relatively smooth hydrological signals; then, by applying Gaussian filters to the model, the scale factors required to remove the leakage signals in the river basin due to weakness during post-processing was estimated. The estimated result shows that the scale factor for the Abbay river basin signal recovery with Gaussian filter at a radius of 150 km is 1.036, 200 km is 1.078, and 300 km is 1.159. From this result, we observe that a higher filtering radius corresponds to more signal leakage in the model. The signal leakage of the model was minimized by the optimal combination of algorithms. Then, monthly cross correlation for the estimated GRACE time series dataset was carried out, and the result indicates CSR + JPL (0.97), CSR + GFZ (0.96), and GFZ + JPL (0.95) for the measured dataset by the three centers (CSR, GFZ, and JPL). All correlation coefficients show a positive correlation greater than 95 percent, indicating that the terrestrial water storage data of the Abbay river basin measured by the three centers were highly consistent.

After calculating the change in equivalent water thickness for the study area, the average water equivalent thickness was estimated, and the result is presented in **Figure 2** below. The result demonstrates the continuous rising and falling of the equivalent water thickness in the Abbay river basin. As shown in **Figure 2**, the highest liquid water equivalent thickness was observed in September 2007 and October 2014 with a value of 14.33 and 14.68 cm per month, respectively, whereas the minimum liquid water equivalent was observed in March 2004 and April 2011 with a value of −10.52 and −10.43 cm per month, respectively. These variations were due to changes in water storage in hydrological reservoirs, which are consistent with similar previous studies [35–37].

3.2 Characteristics of water budget components in Abbay river basin

The characteristics of canopy water storage, soil moisture, surface water storage, surface runoff, evapotranspiration, precipitation, and temperature show a significant variation for the study area. The average value of canopy water storage ranges from 23.05 to 31.23 mm per year for the study period. Planting trees and other vegetation with large water storage capacities helps farmers collect rainfall and lessen runoff, which increases the amount of water that seeps into the ground and replenishes the groundwater aquifer. There are various methods and models to estimate evapotranspiration (EVT) [37–39], but for this study, remote sensing from satellites was used.

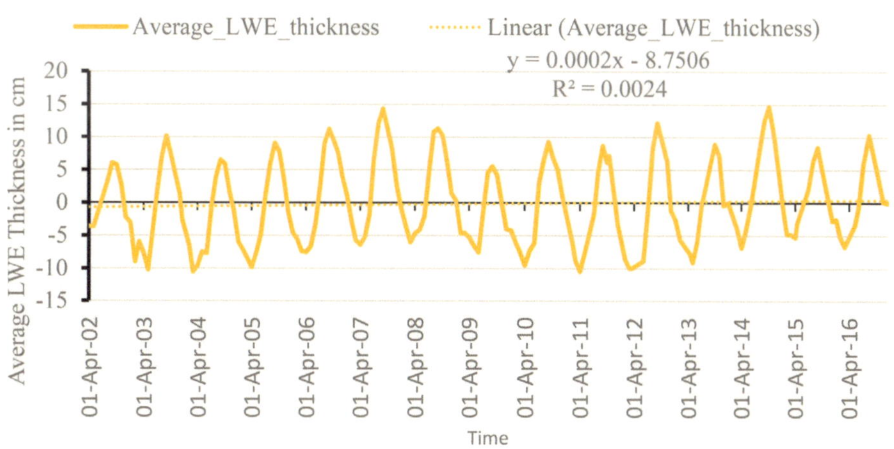

Figure 2.
The average GRACE liquid water equivalent thickness.

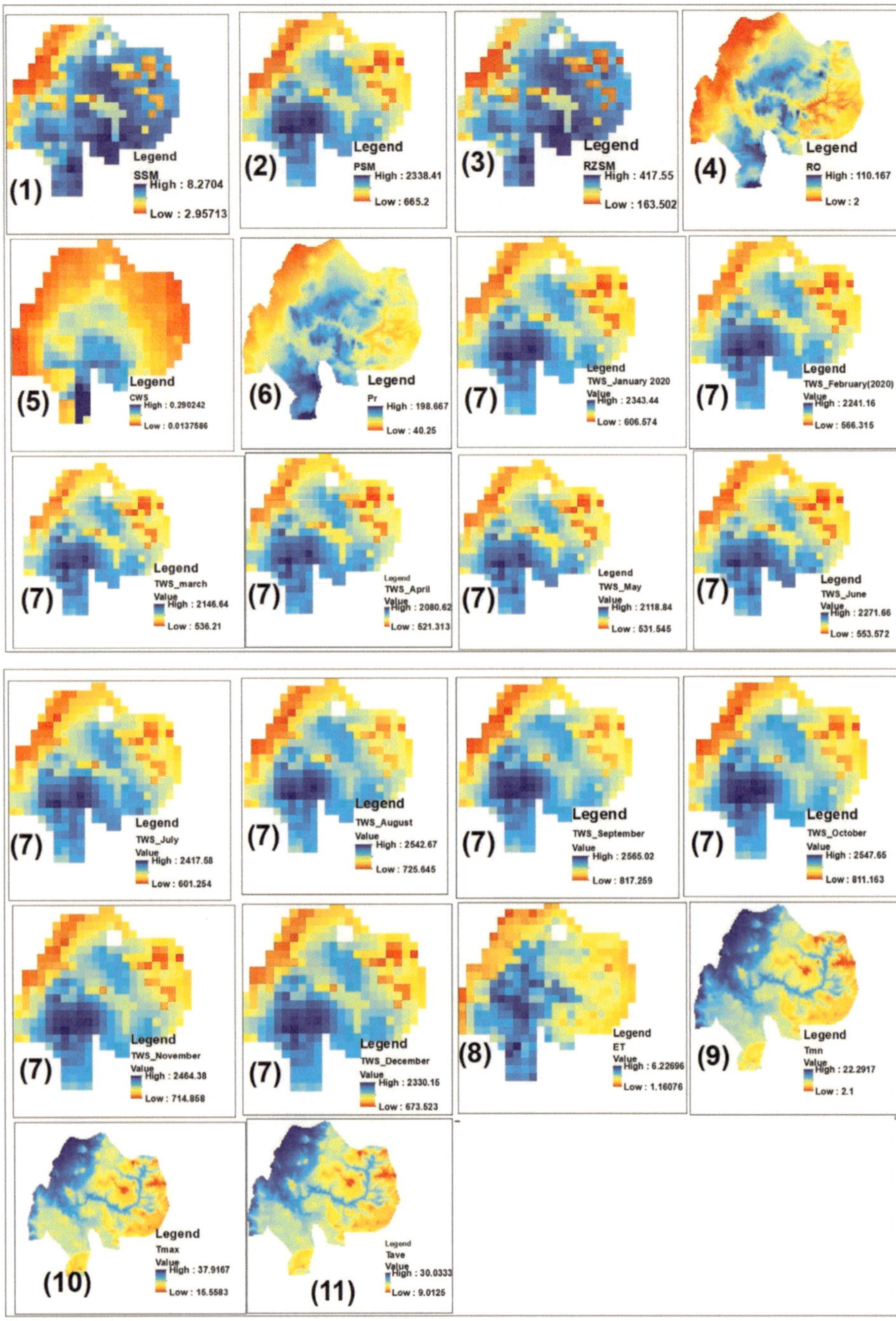

Figure 3.
Water storage components in Abbay river basin: 1. surface soil moisture, 2. profile soil, 3. root zone soil, 4. surface runoff, 5. canopy water storage, 6. precipitation, 7. Terrestrial water storage, 8. Evapotranspiration, 9. Minimum temperature, 10. Maximum temperature, 11. Average temperature.

The analyzed temporal variation of EVT shows a decreasing trend and ranges from 18736.46 to 25588.0 mm per year. Surface runoff in the study area shows a decreasing trend. The Abbay river basin has 49.6 billion cubic meters of surface runoff, according

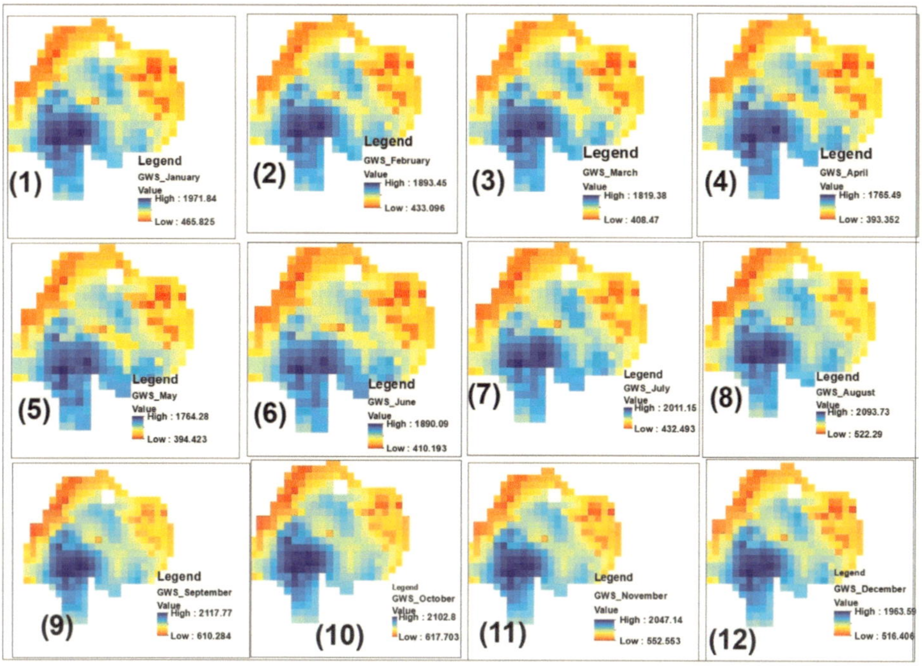

Figure 4.
Monthly spatial distribution of groundwater storage in the Abbay river basin for the year 2020: 1–12 Groundwater storage variation.

to a study done by the Ministry of Water Research, which shows a decreasing trend and ranges from 8145.43 to 14211.19 mm for the study period. However, it is important to note that the surface runoff varies greatly depending on the season and year and influenced by climate, land use, soil type, and topography, which agrees with other similar studies [37–41]. The estimated surface soil moisture, root zone soil moisture, and profile soil moisture show an increasing trend and ranges from 2116.25 to 2308.03 mm, 110207.11 to 117140.84 mm, 521078.64 to 557923.0 mm, respectively. Many variables, including temperature, precipitation, vegetation cover, soil type, and land use, can have a significant impact on soil moisture levels. In order to comprehend the effects of climate change on the ecosystem and human activity, monitoring soil moisture levels is crucial for forestry, agriculture, weather forecasting, and water resource management. The classification of soil moisture was based on the global land data assimilation system [42]. **Figure 3(1–6)** and **Figure 4(1, 2)** show the spatial distribution of the water balance components for the Abbay river basin in 2020. The result displays a significant variation in surface soil moisture, profile soil, root zone soil, surface runoff, canopy water storage, precipitation, evapotranspiration, and temperatures. The variation in components of water storage is due to the extraction of water for domestic water demand, irrigation water supply, and climatic change, which agrees with similar studies [35–39].

3.3 Temporal change of terrestrial water storage for Abbay river basin from GLDAS

For the study area, terrestrial water storage was estimated using GLDAS datasets, which include soil moisture, plant canopy water storage, and surface water storage,

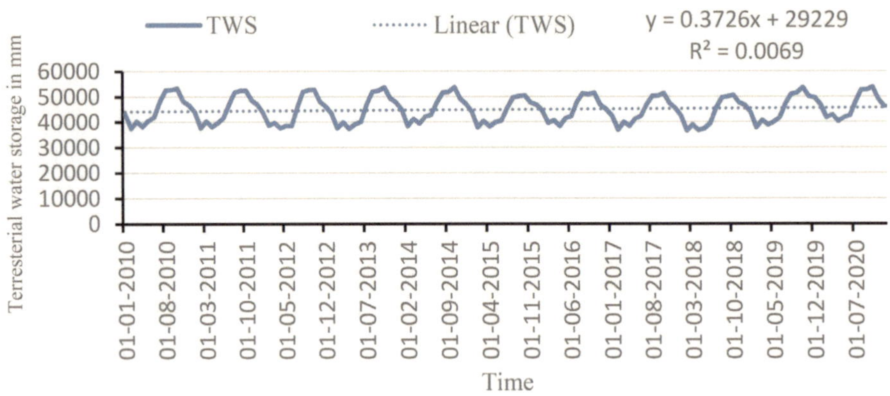

Figure 5.
Year-to-year change in terrestrial water storage for Abbay river basin.

which account for 98%, 0.012, and 1.6% of the river basin, respectively. The variation in terrestrial water storage and year-to-year change in terrestrial water storage for the river basin from GLDAS over the study period show a continuous falling and rising following the seasonal pattern of the study area (**Figure 5**), which is confirmed by a previous study [37–39]. The variation in terrestrial water storage was controlled by seasonal variation and could be due to an even distribution of precipitation, variation of soil and surface water storage, changes in climatic parameters, and drought experience, which is consistent with similar studies [37–39, 43–45].

3.4 Spatiotemporal variation of terrestrial water storage from GLDAS in Abbay river basin

The monthly spatiotemporal variation of the terrestrial water storage shows a significant variation demonstrating the highest spatiotemporal change in terrestrial water storage in September, October, and August. This difference in terrestrial water storage was due to rainfall variation, changes in temperature, land use changes, and different human activities. It can be explained that the study area receives the highest rainfall during September, October, and August, which is confirmed by the precious study [37–39]. As shown in **Figure 3(7)**, the monthly variation of spatiotemporal water storage in 2020 shows a similar trend with minimal alteration. The highest spatial variation of terrestrial water storage was observed in the southwestern parts of the river basin, which indicates slight changes from January to December presented in **Figure 3(7)**. This variation in terrestrial water storage for the study area was due to leakage from irrigation canals, distribution of precipitation, climatic control and leakage from reservoir storage, land use, land cover type, soil type, and others, and this finding agrees with similar previous studies carried out in different areas [37–39, 43–45].

3.5 Temporal variation of groundwater storage from GLDAS

Groundwater storage for the Abbay river basin was estimated by removing surface water storage, soil moisture, and canopy water storage from terrestrial water storage for the study area. The year-to-year variation of groundwater storage was presented in **Figure 6**.

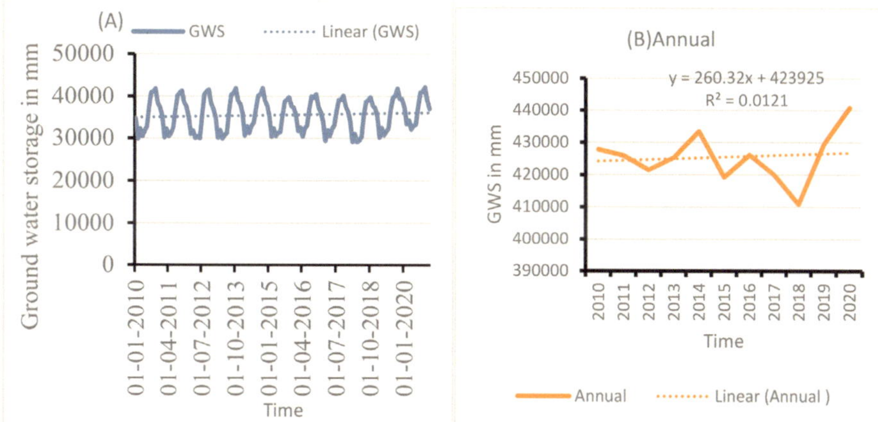

Figure 6.
Characteristics of groundwater storage. (A) interannual groundwater storage. (B) Annual average of groundwater storage in the study area.

As presented in **Figure 3a**, the interannual variation of groundwater storage shows a continuous falling and rising trend, which was controlled by seasonal variation. As presented in **Figure 6B**, the annual groundwater storage variation showed decreasing trends from 2010 to 2012, 2014 to 2015, and 2016 to 2018, whereas there were increasing trends from 2012 to 2014, 2015 to 2016, and 2018 to 2020 in the study area. The estimated groundwater storage of the Abbay river basin ranges from 410871.5 to 440782.2 mm per year during the study period. As seen in **Figure 6**, the minimum annual groundwater storage was observed in 2018, whereas the maximum groundwater storage was observed in 2020. Additionally, the increasing demand for water for irrigation agriculture, water supply, and deforestation agrees with similar previous studies [46–48]. According to statistical data from the "Ethiopia Water Resource Bulletin," the groundwater supply of the country has been steadily increasing on a yearly basis, resulting in a decrease in groundwater in this basin.

The monthly variation of groundwater storage for the Abbay river basin demonstrates a positive trend in January–July, November, and December, whereas it shows negative trends in August, September, and October (**Figure** 7). The fluctuation of groundwater storage was due to excessive withdrawal of water for irrigation, domestic water demands, uneven distribution of precipitation, limited recharge, and climatic changes. During the wet season, high water was infiltrated to join groundwater, whereas in the dry season, water deficit is the main problem in the river basin, and the variation in groundwater storage is due to this condition, which agrees with related studies [48]. We observed that groundwater variation in the Abbay river basin has a strong groundwater storage component and is largely derived from annual precipitation.

3.6 Spatiotemporal characteristics of groundwater storage for Abbay river basin

The spatial variation of the trend in groundwater storage over the Abbay river basin reflects the complexity of hydrological processes and geology in the river basin. The maximum spatiotemporal distribution of groundwater storage was observed in September with a value of 2118 mm, whereas the minimum groundwater storage was observed in April with a value of 393.4 mm (**Figure 4(1-12)**). It is likely that the

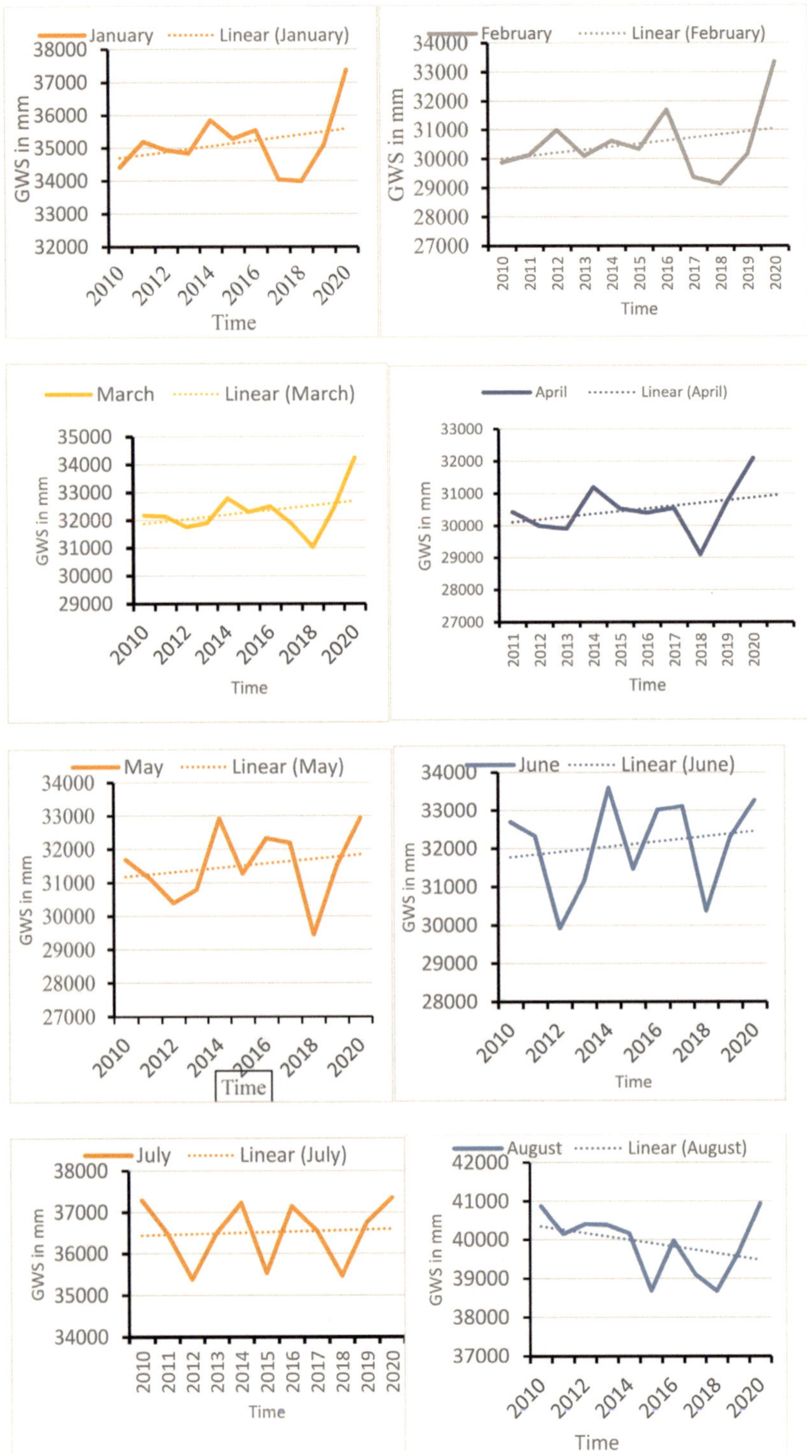

Figure 7.
Monthly temporal trend of groundwater storage in the Abbay river basin from 2010 to 2020.

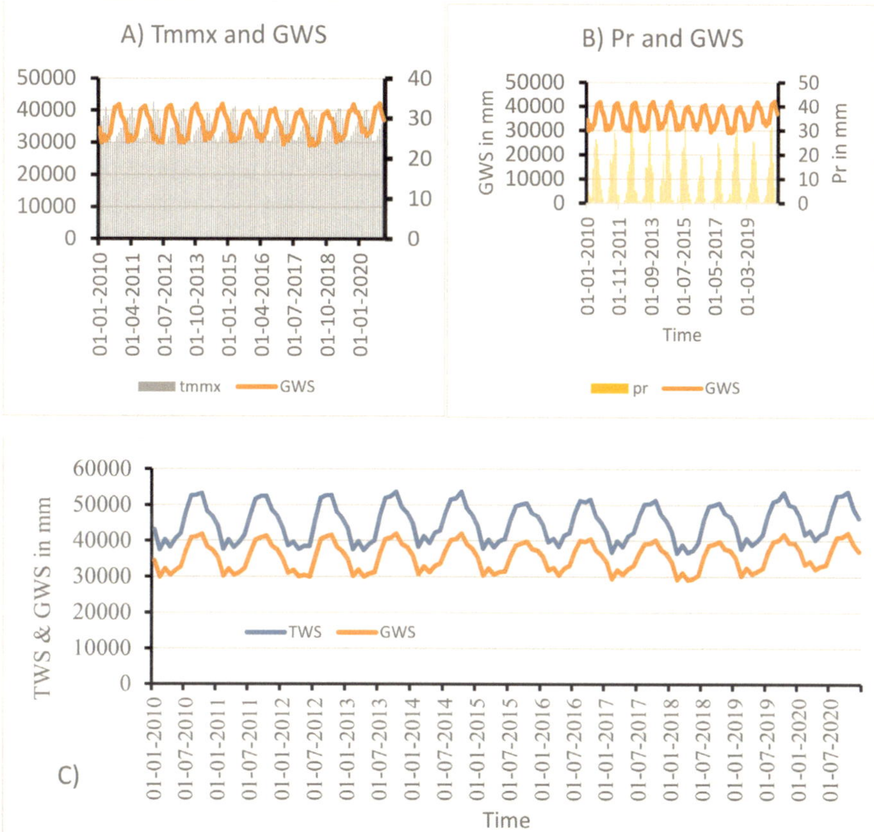

Figure 8.
Characteristics of climatic variable with groundwater storage. (A) Relationship between temperature and groundwater storage. (B) Relationship between precipitation and groundwater storage, (C) Groundwater storage and terrestrial water storage.

groundwater storage trend in the area broadly coincides with terrestrial water storage variation and follows rainfall trends. The climatic change impact on the world in terms of increased or decreased levels of rainfall, water storage, water inundation, and drought, and the water storage in semiarid or arid regions largely depends on precipitation, which affects water storage, as confirmed by similar studies [49–51]. Additionally, land use, land cover, leakage from irrigation canals and irrigation water applications, soil type, and leakage from reservoirs can also lead to variations in groundwater storage for the study area, which coincides with similar studies [52, 53].

3.7 Understanding drivers of groundwater variability

This study presents the spatiotemporal pattern of groundwater storage, precipitation, temperature, and terrestrial water storage. Multiple linear regression analysis is applied to understand the drivers of change in groundwater storage in the study area. The estimated cross-correlation result shows a coefficient of 0.65 for precipitation, 0.99 for terrestrial water storage, and −0.715 for temperature. The result shows that groundwater storage has a high cross-correlation with terrestrial water storage and precipitation, whereas there is a negative cross-correlation with the temperature of the area. The constancy that we observed may be due to interannual changes in rainfall or the complexity of the hydro-geological condition of the study area [54]. The result demonstrates high consistency of groundwater storage with terrestrial water storage, precipitation, and negatively consistent with temperature in the study area. The characteristics of precipitation and temperature with groundwater storage are presented in **Figure 8 (A** and **B)**. As presented in **Figure 8B**, we observed that precipitation has a leading role in driving groundwater variation; we observed high groundwater storage for peak precipitation and low groundwater storage for low precipitation. As shown in **Figure 7a**, we observed high groundwater storage for low temperatures and low groundwater storage for high temperatures, which agrees with a similar study [54]. In addition to precipitation and temperature, other non-climatic factors, such as human water extraction, have also been recognized as driving groundwater storage for the study area, which is confirmed by similar studies [55]. Factors causing variations in groundwater storage may decrease the number of springs and affect groundwater-dependent ecosystems. As shown in **Figure 8C**, groundwater storage and terrestrial water storage show a similar trend.

Land use gives necessary information regarding infiltration, soil moisture, and surface runoff, which affects groundwater occurrence. Areas having good land cover have high infiltration rate and recharge groundwater, whereas areas with low land use land cover demonstrate low infiltration rate and experiencing less recharge to groundwater, which was consistent with other studies [8].

Irrigated agriculture is the largest consumer of groundwater resources. For more efficient and sustainable utilization of the limited water resources, an improved understanding of how to respond to irrigation is essential. The interaction between agricultural irrigation and groundwater resources is generally very close to each other. The excess irrigation water extracted in the river basin leads to groundwater depletion, whereas recharges leads to increase in groundwater storage and consistent with other similar studies [56].

The use of satellite data can improve the applicability of data-scarce areas with limited hydrological variables, such as groundwater observation well data with uneven spatial and temporal distribution. However, these data sources also bring significant uncertainty. For example, GRACE data have their own uncertainty in data processing, and other input data have contributed to uncertainty and potential error propagation of the total input data. GRACE also observes TWS depletion associated with anthropogenic groundwater extraction. The model, however, does not represent anthropogenic groundwater withdrawals and is not skilled in reproducing the interannual variability of groundwater. Assimilation of GRACE TWS introduces long-term trends and improves the interannual variability in groundwater. But the assimilation also introduces a negative trend in simulated evapotranspiration, whereas in reality, evapotranspiration is likely enhanced by irrigation, which is also unmodeled, which agrees with similar finding [57].

4. Conclusion

In this study, the ways to process NASA's GRACE and GLDAS satellite mission data to understand dynamic changes in water balance components were fixed for the study area, and the spatiotemporal variation in groundwater storage of the Abbay river basin during the time period of 2010–2020 was estimated by using the GLDAS and GRACE-driven Naoh model outputs. The characteristics of water budget components in the Abbay river basin were quantified, the sensitivity of each variable for the contribution of terrestrial water storage and groundwater storage was evaluated, and the impacts of climatic variables were further analyzed through the variation in groundwater storage in the study area.

From the point of view of interannual variation, the estimated LWE from GRACE and TWS from GLDAS show the same trends during the study period. To quantify terrestrial water storage, correlation between the selected variables was correlated using R-studio, and the correlated result shows that a random forest model is the best to predict variables, and the terrestrial water storage was quantified. The result demonstrates a continuous falling and rising interannually. The spatial variation of the groundwater storage in the river basin shows a similar trend with little variation. We observed high groundwater storage in the southwestern parts of the river basin. The mean annual groundwater storage decreased from 2010 to 2012, 2014 to 2015, and 2016 to 2018, while increasing from 2012 to 2014, 2015 to 2016, and 2018 to 2020.

Climate change and anthropogenic activities remarkably influence water resource systems. The result obtained here indicates that groundwater storage is easily affected by precipitation and temperature, both directly and indirectly. The correlation coefficient shows a negative correlation with temperature and a positive correlation with precipitation. Additionally, we provide evidence that GRACE and GLDAS provide users with valid datasets even in areas with little observation. The quantitative outcome demonstrates that long-term monitoring of water storage is necessary to avoid potential future water shortages and to conserve natural resources. The GRACE and GLDAS observations have recently been identified as effective tools in areas with a lack of direct *in situ* measurements and large-scale basins. The result of the study also shows that groundwater storage depletion is likely to increase in the future due to the development of traditional irrigation practices across the basin. It will have adverse consequences for water and food security in the river basin. It is recommended for policymakers and stakeholders to improve and enhance direct water level measurement and increase recharging for better sustainable management of groundwater resources in the river basin. It is also of vital importance in terms of saving freshwater and delivering it to future generations.

Author details

Agegnehu Kitanbo Yoshe[1,2]

1 Department of Water Resources and Irrigation Engineering, Arba Minch University, Arba Minch, Ethiopia

2 Department of hydrogeology, Siberian School of Geosciences, Irkutsk National Research Technical University, Irkutsk, Siberia, Russia

*Address all correspondence to: kitanbo@gmail.com

References

[1] Pande CB. Watershed management and development. In: Sustainable Watershed Development. Springer Briefs in Water Science and Technology. Cham: Springer; 2020b. DOI: 10.1007/978-3-030-47244-3_2

[2] Ali S, Wang Q, Liu D, Fu Q, Mafuzur Rahaman M, Abrar Faiz M, et al. Estimation of spatio-temporal groundwater storage variations in the lower transboundary Indus Basin using GRACE satellite. Journal of Hydrology. 2022;**605**:127315. DOI: 10.1016/j.jhydrol.2021.127315

[3] Pande CB. Introduction. In Sustainable Watershed Development. Springer Briefs in Water Science and Technology. Cham: Springer; 2020a. DOI: 10.1007/978-3-030-47244-3_1

[4] Elliott J, Deryng D, Müller C, Frieler K, Konzmann M, Gerten D, et al. Constraints and potentials of future irrigation water availability on agricultural production under climate change. Proceedings of the National Academy of Sciences. 2014;**111**(9):3239-3244. DOI: 10.1073/pnas.1222474110

[5] Berhanu B, Seleshi Y, Melesse AM. Surface water and groundwater resources of Ethiopia: Potentials and challenges of water resources development. In: Melesse A, Abtew W, Setegn S, editors. Nile River Basin. Cham: Springer; 2014. pp. 97-117

[6] Jha MK, Chowdary V, Chowdhury A. Groundwater assessment in Salboni block, West Bengal (India) using remote sensing, geographical information system and multi-criteria decision analysis techniques. Hydrogeology Journal. 2010;**18**:1713-1728

[7] Richey AS, Thomas BF, Lo MH, Reager JT, Famiglietti JS, Voss K, et al. Quantifying renewable groundwater stress with GRACE. Water Resources Research. 2015;**51**:5217-5238

[8] Yoshe AK. Integrated approach for groundwater potential exploration in Abbay River Basin, East Africa. Sustainable Water Resources Management. 2024;**10**:93. DOI: 10.1007/s40899-023-01026-7

[9] Khan S, Rana T, Gabriel HF, Ullah MK. Hydrogeologic assessment of escalating groundwater exploitation in the Indus Basin. Pakistan Hydrogeology Journal. 2008;**16**:1635-1654. DOI: 10.1007/s10040-008-0336-8

[10] Cheema MJM, Immerzeel WW, Bastiaanssen WGM. Spatial quantifcation of groundwater abstraction in the irrigated indus basin. Groundwater. 2014;**52**:25-36. DOI: 10.1111/gwat.12027

[11] Fashae OA, Tijani MN, Talabi AO, Adedeji OI. Delineation of groundwater potential zones in the crystalline basement terrain of SW-Nigeria: An integrated GIS and remote sensing approach. Applied Water Science. 2014;**4**:19-38

[12] Anteneh ZS, Awoke BG, Reda TM, Jothimani M. Groundwater potential mapping using integrations of remote sensing and analytical hierarchy process methods in Ataye-watershed, middle Awash Basin, Ethiopia. Sustainable Water Resources Management. 2022;**8**:183. DOI: 10.1007/s40899-022-00772-4

[13] Oh H-J, Kim Y-S, Choi J-K, Park E, Lee S. GIS mapping of regional probabilistic groundwater potential in the area of Pohang City, Korea. Journal of Hydrology. 2011;**399**:158-172

[14] Kaliraj S, Chandrasekar N, Magesh N. Identification of potential groundwater recharge zones in Vaigai upper basin, Tamil Nadu, using GIS-based analytical hierarchical process (AHP) technique. Arabian Journal of Geosciences. 2014;7:1385-1401

[15] Pourtaghi ZS, Pourghasemi HR. GIS-based groundwater spring potential assessment and mapping in the Birjand township, southern Khorasan Province, Iran. Hydrogeology Journal. 2014;**22**: 643-662

[16] Nampak H, Pradhan B, Manap MA. Application of GIS based data driven evidential belief function model to predict groundwater potential zonation. Journal of Hydrology. 2014;**513**:283-300

[17] Batelaan O, De SF. WetSpass: a flexible, GIS based, distributed recharge methodology for regional groundwater modelling, research gate; 2001, impact of human activity on groundwater dynamics. In: Proceedings of a Symposium Held During the Sixth IAHS Scientific Assembly at Maastricht. The Netherlands: IAHS Publ; 2001

[18] Sun AY. Predicting groundwater level changes using GRACE data. Water Resources Research. 2013;**49**:5900-5912

[19] Swenson SC. GRACE Monthly Land Water Mass Grids NETCDF RELEASE 5.0. Ver. 5.0. PO.DAAC, CA, USA2012. Dataset accessed [YYYY-MM-DD] at. DOI: 10.5067/TELND-NC005

[20] Hussein AA, Govindu V, Nigusse AGM. Evaluation of groundwater potential using geospatial techniques. Applied Water Science. 2016;7:2447-2461. DOI: 10.1007/s13201-016-0433-0

[21] Velis M, Conti KI, Biermann F. Groundwater and human development: Synergies and trade-offs within the context of the sustainable development goals. Sustainability Science. 2017;**12**: 1007-1017. DOI: 10.1007/s11625-017-0490-9

[22] Mukherji A. Sustainable groundwater Management in India Needs a water-energy-food nexus approach. Applied Economic, Perspectives and Policy, Special Issue: CWAE 40th Anniversary. 2022;**44**(1): 394-410. DOI: 10.1002/aepp.13123

[23] Mehdi A, Mobin E, Mohammad A, Elyasi AH, Zahra N. Application assessment of GRACE and CHIRPS data in the Google earth engine to investigate their relation with groundwater resource changes (Northwestern region of Iran). Journal of Groundwater Science and Engineering. 2023;**9**(2):102-113. DOI: 10.19637/j.cnki.2305-7068.2021.02.002

[24] Kumar D, Bhattacharjya RK. GRNN model for prediction of groundwater fluctuation in the state of Uttarakhand of India using GRACE data under limited bore well data. Journal of Hydroinformatics. 2021;**23**(3):567-588. DOI: 10.2166/hydro.2021.108

[25] Tapley BD, Bettadpur S, Ries JC, Thompson PF, Watkins MM. GRACE measurements of mass variability in the earth system. Science. 2004;**305**:503-505

[26] Case K, Kruizinga G, Wu S. GRACE Level 1B Data Product User Handbook. Jet Propulsion Laboratory; 2010

[27] Landerer FW, Swenson SC. Accuracy of scaled GRACE terrestrial water storage estimates. In: Advancing Earth and Space Sciences. Vol. 48. Water Resources Research; 2012. p. W04531, 11 PP. DOI: 10.1029/2011WR011453

[28] Ramillien G, Frappart F, Güntner A, Ngo-Duc T, Cazenave A, Laval K. Time variations of the regional

evapotranspiration rate from gravity recovery and climate experiment (GRACE) satellite gravimetry. Water Resources Research. 2006;**42**:10403

[29] Rodell M, Houser P, Jambor U, Gottschalck J, Mitchell K, Meng C, et al. The global land data assimilation system. Bulletin of the American Meteorological Society. 2004;**85**:381-394

[30] Karami GH, Bagheri R, Rahimi F. Determining the groundwater potential recharge zone and karst springs catchment area: Saldoran region, western Iran. Hydrogeology Journal. 2016;**24**:1981-1992. DOI: 10.1007/s10040-016-1458-z

[31] Harris I, Jones PD, Osborn TJ, Lister DH. Climatic Research Unit2017. DOI: 10.5285/edf8febfdaad48abb2cbaf7d7e846a86

[32] Sridhar V, Wedin DA. Hydrological behavior of grasslands of the Sandhills: Water and energy balance assessment from measurements, treatments and Modeling. Ecohydrology. 2009;**2**:195-212

[33] Rahaman MM, Thakur B, Kalra A, Ahmad S. Modeling of GRACE-derived groundwater information in the Colorado River Basin. Hydrology. 2019;**6**:19

[34] Tirkey N, Parhi PK, Lohani AK, Chandniha SK. Analysis of precipitation variability over Satluj Basin, Himachal Pradesh, India: 1901–2013. Journal of Water and Climate Change. 2020;**12**(7):jwc2020136

[35] Chen L, He Q, Liu K, Li J, Jing C. Downscaling of GRACE-derived groundwater storage based on the random forest model. Remote Sensing. 2019;**11**:2979

[36] Moghim S. Assessment of water storage changes using GRACE and GLDAS. Water Resources Management. 2020;**34**(2):685-697. DOI: 10.1007/s11269-019-02468-5

[37] Yoshe AK. Estimation of change in terrestrial water storage for Abbay River Basin, Ethiopia. Hydrology Research. 2023;**54**(11):1451. DOI: 10.2166/nh.2023.119

[38] Yoshe AK. Water availability identification from GRACE dataset and GLDAS hydrological model over data-scarce river basins of Ethiopia. Hydrological Sciences Journal. 2024;**69**(6):721-745. DOI: 10.1080/02626667.2024.2333852

[39] Yoshe AK. Assessment of anthropogenic and climate-driven water storage variations over water-stressed river basins of Ethiopia. Hydrology Research. 2024;**55**(3):351. DOI: 10.2166/nh.2024.169

[40] Chaemiso S, Kartha S, Pingale S. Effect of land use/land cover changes on surface water availability in the Omo-gibe basin, Ethiopia. Hydrological Sciences Journal. 2021;**66**(13):1936-1962

[41] Yenehun A, Walraevens K, Batelaan O. Spatial and temporal variability of groundwater recharge in Geba basin, northern Ethiopia. Journal of African Earth Sciences. 2017;**134**:198-212

[42] Li B, Rodell M, Kumar S, Beaudoing HK, Getirana A, Zaitchik BF, et al. Global GRACE data assimilation for groundwater and drought monitoring: Advances and challenges. Water Resources Research. 2019;**55**:7564-7586

[43] Almazroui M. Assessment of meteorological droughts over Saudi Arabia using surface rainfall observations during the period 1978–2017. Arabian Journal of Geosciences.

2019;**12**:694. DOI: 10.1007/s12517-019-4866-2

[44] Abdollahi K, Bashir I, Verbeiren B, Harouna MR, Van Griensven A, Huysmans M, et al. A distributed monthly water balance model: Formulation and application on Black Volta Basin. Environment and Earth Science. 2017;**76**(5):198

[45] Yisehak B, Shiferaw H, Abrha H, Gebremedhin A, Hagos H, Adhana K, et al. Spatiotemporal characteristics of meteorological drought under changing climate in semi-arid region of northern Ethiopia. Environmental Systems Research. 2021;**10**:21. DOI: 10.1186/s40068-021-00226-4

[46] Xiao R, He X, Zhang Y, Ferreira V, Chang L. Monitoring groundwater variations from satellite gravimetry and hydrological models: A comparison with in-situ measurements in the mid-Atlantic region of the United States. Remote Sensing. 2015;7:686-703

[47] Tiwari VM, Wahr J, Swenson S. Dwindling groundwater resources in northern India, from satellite gravity observations. Geophysical Research Letters. 2009;**36**(18):P1-6, L18401. DOI: 10.1029/2009GL039401.

[48] Feng W, Wang C, Mu D, Zhong M, Zhong Y, Xu H. Groundwater storage variations in the North China plain from GRACE with spatial constraints. Chinese Journal of Geophysics. 2017;**60**:1630-1642

[49] Mondal A, Khare D, Kundu S. Change in rainfall erosivity in the past and future due to climate change in the central part of India. International Soil and Water Conservation Research. 2016; **4**:186-194

[50] Ndehedehe CE, Ferreira VG, Agutu NO, Onojeghuo AO, Okwuashi O, Kassahun HT, et al. What if the rains do not come? Journal of Hydrology. 2021; **595**:126040

[51] Prein AF, Liu C, Ikeda K, Trier SB, Rasmussen RM, Holland GJ, et al. Increased rainfall volume from future convective storms in the US. Nature Climate Change. 2017;7:880-884

[52] Bhanja SN, Mukherjee A, Rodell M, Wada Y, Chattopadhyay S, Velicogna I, et al. Groundwater rejuvenation in parts of India influenced by water-policy change implementation. Scientific Reports. 2017;7:7453

[53] Thomas BF, Famiglietti JS. Identifying climate-induced groundwater depletion in GRACE observations. Scientific Reports. 2019;**9**: 4124

[54] Ordens CM, McIntyre N, Underschultz JR, Ransley T, Moore C, Mallants D. Preface: Advances in hydrogeologic understanding of Australia's great Artesian Basin. Hydrogeology Journal. 2020;**28**:1-11

[55] Ndehedehe CE, Ferreira VG. Assessing land water storage dynamics over South America. Journal of Hydrology. 2020;**580**:124339

[56] Fang Q, Chen Y, Yu Q, Ouyang Z, Li Q, Yu S. Much improved irrigation use efficiency in an intensive wheat-maize double cropping system in the North China plain. Journal of Integrative Plant Biology. 2007;**49**:1517-1526. DOI: 10.1111/j.1672-9072.2007.00559.x

[57] Girotto M, De Lannoy GJM, Reichle RH, Rodell M, Draper C, Bhanja SN, et al. Benefits and pitfalls of GRACE data assimilation: A case study of terrestrial water storage depletion in India: India GRACE data assimilation. Geophysical Research Letters. 2017;**44**(9). DOI: 10.1002/2017GL072994

Chapter 3

Delineation of Recharge Zones for Sustainable Protection of Groundwater in Bidou I (South Cameroon)

Olivier Njikeu, Roger Feumba and Benjamin Ngounou Ngatcha

Abstract

The aim of this work is to delineate groundwater recharge potential zones in Bidou I, covering approximately a surface area of 144 km^2. To meet this goal, seven influencing parameters were merged in a GIS. A total of 2370 Lineaments have been extracted and mapped using remote sensing and GIS technics. Their lengths vary between 60 and 4498 m with an average of 525 ± 420 m. Eight deformed rock types and two major fractures cross-cutting each other were identified. The mapped lineaments main directions (N330°N–045°E) define the groundwater flowing direction corresponding to the recharge average direction (N010°E) of the aquifers in the area. The water balance calculation shows that about 10% of the average interannual rainfall in the Kribi region (1991–2021) was effectively infiltrated. This implies that the effective recharge feeds the stock variation (ΔRi). The calculated rock filtration velocities range from 0.5×10^{-9} to 11.0×10^{-9} m/s with an average of 2.5×10^{-9} m/s. The LULC classes expressed as percentages of the total study area are: Vegetation: 90%, Water bodies: 5%, Bare soils: 4% and Built-up: 1%. In prelude to sustainable groundwater protection and management in the area, a five classes map (very poor, poor, moderate, good and very good) of Groundwater Recharge Potential Zones (GWRPZ) was realized by assigning to each layer a fixed weight and a score from Analytical Hierarchy Process (AHP) using weighted overlay analysis (WOA) tool.

Keywords: remote sensing, AHP, WOA, groundwater recharge potential zones, sustainable protection, Bidou I (South Cameroon)

1. Introduction

Lack of awareness, scarcity and poor management of water resources in the world, particularly in Africa's arid zones is becoming increasingly widespread, even in areas with high rainfall [1–4]. The groundwater recharge is directly affected by the rainfall amount, and a significant change in rainfall can cause a similar effect in it [5–7]. However, rainfall patterns vary considerably as a result of climate change. Intense precipitation concentrated over a short period leads to devastating floods, while late and scanty rainfall causes drought conditions, resulting in reduced recharge [8]. Given the potential negative effects of global climate change on water resources, the population is obligated to focus on groundwater supply for agricultural and food purposes [9, 10]. Thus, evaluating the groundwater recharge potential is very important for secured groundwater systems. It is therefore imperative to identify appropriate groundwater recharge zones in order to address groundwater protection and guarantee the long-term viability of aquifers [11]. Many studies on the factors affecting groundwater recharge have been already done to clear up the groundwater hydrogeology by calculating the groundwater recharge potential index using the Analytical Hierarchy Process (AHP) approach and GIS techniques [11–15]. It is therefore necessary to study the relationship between these influencing factors for the best understanding of their influence on groundwater recharge potential. This study aims to delineate groundwater recharge potential zones in Bidou I by providing a comprehensive understanding of groundwater recharge dynamics, in order to contribute to sustainable water resource protection and management to facilitate informed decision-making processes.

2. Study area

2.1 Geographic and administrative localization

The study area of more than 140 km^2 is situated in the South Cameroon Region in the Ocean Division at about 45 km northeast of the Deep Sea Port of Kribi (**Figure 1**). Belonging to the district of Fifinda the area is limited by Latitudes 2°53′26″–3°03′00″ N and Longitudes 10°03′31″ and 10°09′39″ E. The area's steep relief (≤ 400 m altitude) is dominated by a major shear zone between 200 and 400 m altitude, forming a ridge with a steep topographic slope reaching 200 meters in height difference covered by a moist evergreen and semi-deciduous equatorial forest. Low-lying areas (altitude <100 m) are occupied by marshy valleys. The area is characterized by a humid bimodal climate with two rainy seasons and two dry seasons unevenly distributed. From 1991 to 2021, the average interannual precipitation is 2957 mm, and the average interannual air temperature is 25.67°C. The permanent river network flows into river kienké which is the main river in the area.

2.2 Hydrogeology

The basement formations in Cameroon represent more than 90% of the territory whose aquifers are linked only to fissuring and alteration [4–6, 16–18]. Upon these basements, there are Palaeozoic volcano-sedimentary formations, from the Secondary, the Tertiary and the Quaternary, all topped by Quaternary alluviums that are sometimes very thick (over 50 m), such as that of Douala [1, 16], Benoué [19] and even Grand Yaéré [5, 6]. Groundwater represents approximately 98% of global freshwater resources being therefore the main source of drinking water for the population [5], with an estimated reserve between 667 and 4810 km^3 [20]. However, most of the groundwater appears to be unevenly distributed throughout the country. In the far north, these include the sedimentary formations of the Chadian basin, the Ngodeni collapse trough and the fluvio-lacustrine series, as well as the Quaternary, Tertiary and Precambrian aquifers [5, 6] while in the south the horst of the Mount Cameroun—Bioko volcanic axis isolates the Douala–Kribi—Campo sedimentary basins and the Rio del Rey basin [1, 16]. On the hydrogeological, hydrodynamic and hydrochemical points of view, numerous studies

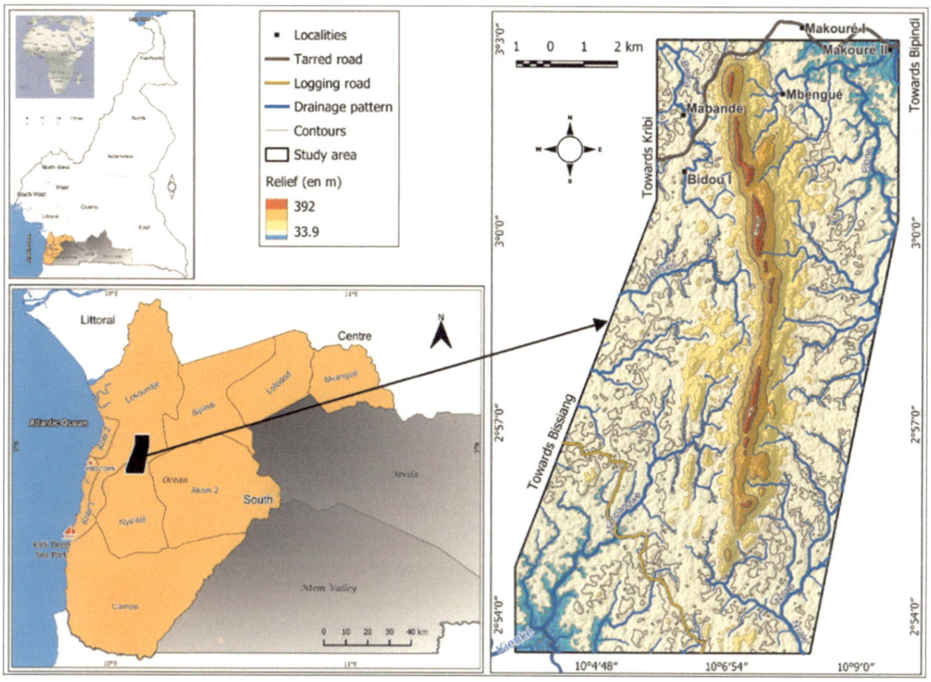

Figure 1.
Localization of the study area.

characterizing water resources have already been carried out in the northern part [5, 6, 21–25]. While in the southern part, hydrogeological studies are few [9, 26–29]. No detailed hydrogeological study about the assessment of the groundwater potential of the whole country has yet been carried out [17]. The same author states that according to [30], a study was carried out on the basis of pre-existing data by BCEOM, SOGREAH and ORSTOM using a balance sheet approach [31–33]. Basement zone in Cameroon is mostly characterized by two types of aquifers sometime superimposed or isolated. They are Laterite aquifers, which are continuous and more or less exploitable, and discontinuous artesian fissured or fractured aquifers. It is a bilayer aquifer when they are superimposed [27–29]. The altered zone is essentially capacitive, whereas the fractured zone is much more conductive. But both functions (capacitive and conductive) coexist within each aquifer [34, 35]. The potential of these aquifers depends on the parameters of the water balance and their configuration [4]. The formations encountered consist of plutonic or metamorphic crystalline rocks, sometimes associated with volcano-sedimentary formations and "greenstone belts" of Precambrian to Cambrian age, characterized by very low porosity, generally less than 1% [27–29]. However, the initial work of [4] in this greenfield area, presents the area with good groundwater potential and good productivity according to their pumping rates [36].

2.3 Geology

2.3.1 Regional geology

Two major litho-structural entities defined the Cameroonian Precambrian basement complex: (1) the North Equatorial Pan-African Fold Belt which covers the

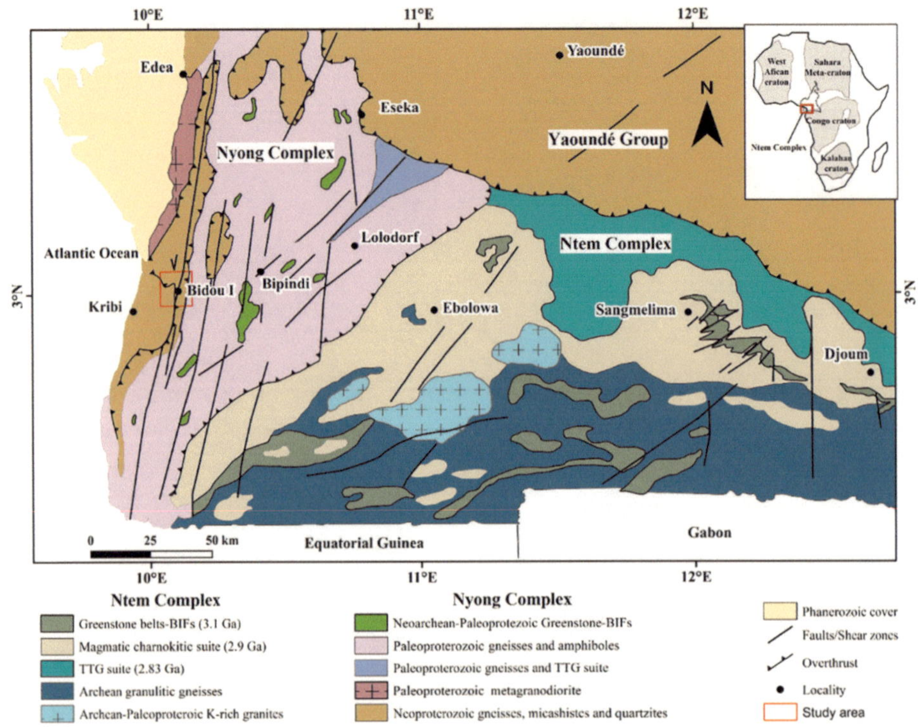

Figure 2.
Study area over the geological map of SW Cameroon (after [4, 17]).

greatest part of the territory from the series of Yaoundé to the Far-North [37, 38] and (2) the cratonic domain, which covers the northern part of the Congo craton [39–43]. The cratonic domain is made up of charnockitic rocks, doleritic dykes and potassic granites [40, 43] which intrude the greenstone belt in the Ntem Complex, made up of three subcomplex (**Figure 2**): Ayina, Ntem and Nyong which covers the study area. The works of [39, 44–46] in this last complex show that it represents a reactivated part of the Archean Congo craton throughout the Eburnean/Transamazonian and Pan-African/Brazilian. The Nyong complex is the Paleoproterozoic contact zone coming from the tectonic event between the Congo and São Francisco cratons [20, 47–50]. It is made up of Archean materials of volcano-sedimentary origin [42, 50–53] grouping gneiss clustered with Banded Iron Formations (BIFs), plutonites and green rocks.

2.3.2 Local geology

The Bidou I area is a zone that has been little explored [4]. The most actual accomplished studies were done by [54, 55] after those of [39]. The outcome shows that the source rock of the metasedimentary rocks of Bidou I were deposited in an active or passive continental margin. They are composed of garnetiferous amphibolitic gneisses and two-micas schists, garnet-kyanite-staurolite (both separated by a regional fault), micas-rich quartzites and BIFs (**Figure 3**). According to

[48], three phases of tectonic events related to the Kribi-Campo shear zone (NNE-SSW) hit the Kribi area and its surroundings, even Bidou I. Subsequently, the works of [56, 57] reveal that Bidou I is made of gneiss, amphibolites, orthogneiss and BIFs weakly metamorphosed and very magnetic with centimetric bands.

3. Materials and methods

3.1 Data source

The data used are: (1) climatic data of the Kribi area (precipitations and temperatures), corresponding to the weather forecasts from 1991 to 2021, downloaded using the link https://fr.climate-data.org/; (2) Cameroon's litho-structural map [48]; (3) geological map of south-west Cameroon [39]; (4) a 1:200,000 scale topographic base maps of Kribi and Edéa; (5) rock types and structural measurements from the field data (6) Satellite images ((i) Landsat 8 OLI_TIRS Collection 2 level 2 (LC08_L2SP) free of cloud (Path_186/Row_058) available on March 25th, 2015, and (ii) SRTM (Shuttle Radar Topography Mission (2014)) with ground resolution of 30 m) freely downloaded on http://earthexplorer.usgs.gov in Universal Transverse Mercator. The LC08_L2SP image used is described in **Table 1**. Slope, drainage pattern, lineament, geomorphology, colored composite and land use/land cover maps were elaborated by processing satellite data. We used geologist hammer, compass and clinometer for rock sample collection and field structural measurements. After this, all the data were integrated for maps production and were realized after and intensive field mapping work. Excel, ArcGIS 10.8, Qgis 3.34, ENVI 5.3, PC geomatica 2016 and Rockworks are the software used in this study (**Figure 3**).

Bands	Description	Wavelength	Resolution
Band 1	Ultra Blue	443.0 nm	30 m
Band 2	Blue	482.0 nm	30 m
Band 3	Green	561.5 nm	30 m
Band 4	Red	654.5 nm	30 m
Band 5	Near Infrared (NIR)	865.0 nm	30 m
Band 6	Shortwave Infrared (SWIR) 1	1608.5 nm	30 m
Band 7	Shortwave Infrared (SWIR) 2	2200.5 nm	30 m
Band 10	Thermal Infrared (TIRS) 1	10895.0 nm	30 m [1]
Band QA	Quality Assessment band	No data	30 m
Band QA_RADSAT	Radiometric Saturation and Terrain Occlusion QA Band	No data	30 m
Band SR_QA_AEROSOL	SR Aerosol QA	No data	30 m

Table 1.
Description of the LC08_L2SP image used.

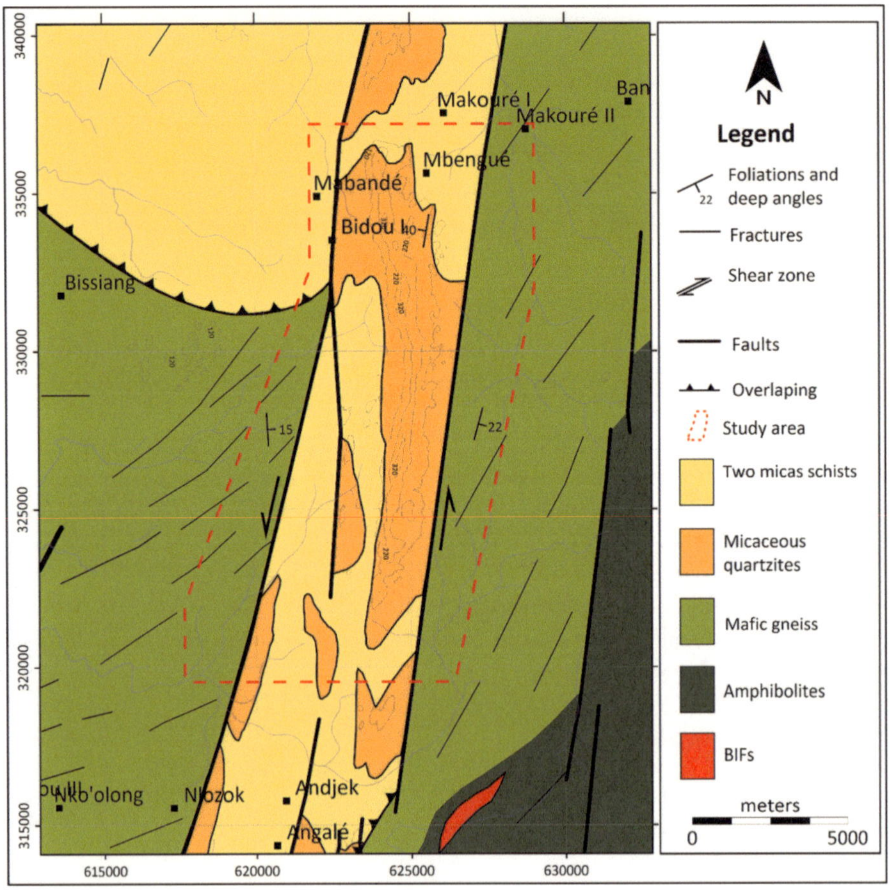

Figure 3.
Geology of the study area, extracted from the geological map of South-West Cameroon (after [4]).

3.2 Methodology

3.2.1 Data pre-processing

In order to enhance the reflectance of features present on the ground and easily detect lineaments, satellite images are supposed to be pre-processed before their utilization. The choice of a method depends on the experience and preferences of the image analyzer, the type of data used and its application. **Figure 4** shows the flow-chart of the methodology applied to the digital data used in this work.

3.2.2 Data processing

3.2.2.1 Pansharpening

Pansharpening is the use of a single panchromatic band (higher spatial resolution) to increase the spatial resolution of the bands of a multispectral image (higher spectral resolution) [58–60]. The result (**Figure 5**) of this task is an image with high spectral resolution and high spatial resolution. After the layers stacking and the contrast

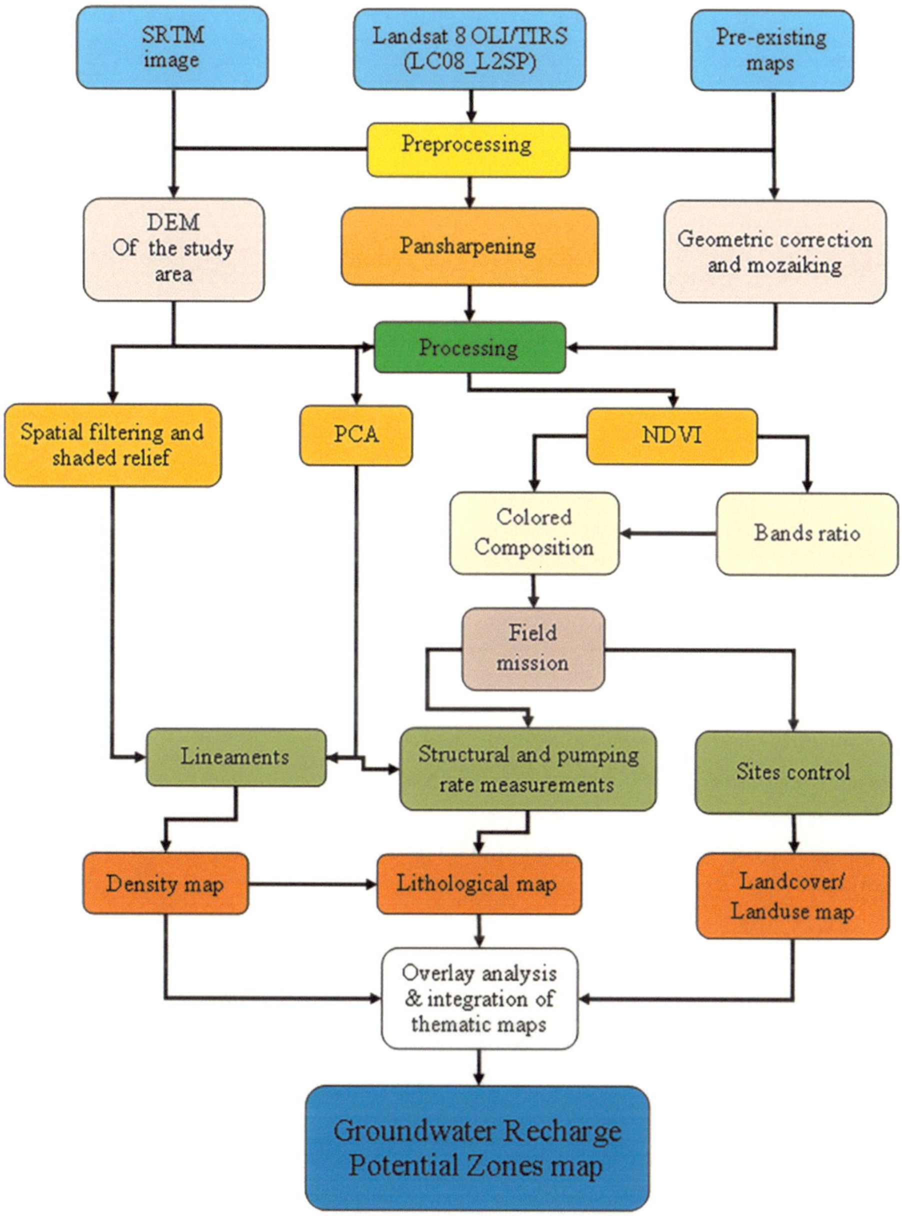

Figure 4.
Flowchart for mapping groundwater recharge potential zones in this work.

enhancement of bands, they were exported in grayscale (8-bit) and filtered. We also performed correlation of these bands and the outcome (**Table 2**) almost shows their common source, implying they are ready for Principal Component Analysis (PCA).

3.2.2.2 Principal component analysis (PCA)

PCA is a computation method used to convert correlated parameters into uncorrelated parameters named principal components [59]. It has been widely

Bandes	Band 2	Band 3	Band 4	Band 5	Band 6	Band 7
Band 2	1					
Band 3	0.6265	1				
Band 4	0.5993	0.8487	1			
Band 5	0.4273	0.5589	0.2013	1		
Band 6	0.7397	0.7377	0.5928	0.725	1	
Band 7	0.7684	0.7411	0.7541	0.4529	0.9127	1

Table 2.
Correlation of the raw bands used (LC08_L2SP).

used in lineaments extraction, statistics and signal processing [61, 62]. To defeat the superfluity and to ameliorate the gap in the image, PCA was applied. For, the ENVI software was used to produce principal components images (PC1, PC2 ... etc.) and 93% of the results were given by PC1 after a mathematical computation from the covariance table (Eq. (1)), making it to be used for the next stages of transformation.

$$\begin{vmatrix} V_1 & 0\ldots & V_1 \\ 0 & V_2\ldots & 0 \\ 0 & 0\ldots & V_N \end{vmatrix} V = E_e^{Cov(E_e)} \tag{1}$$

Where: Cov is the matrix function, E_e eigenvectors matrix, T is the function of transposition, and V is the eigenvalues diagonal matrix (zero for elements out of the diagonal).

3.2.2.3 Water balance estimation

Based on hydroclimatic data of the area of Kribi (1991–2021), the hydrometric parameters of Bidou I were calculated using the method of [63] (Eq. (2)).

$$\mathrm{P} = \mathrm{ETR} + \mathrm{I} + \mathrm{R} + \Delta\mathrm{Ri} \tag{2}$$

Where: ETR = evapotranspiration; I = infiltration, R = runoff, ΔRi = underground water variable reserve and P = average interannual precipitation.

$$ETP = 16\left(\frac{10\mathrm{T}}{\mathrm{I}}\right)^{\alpha} \times F(\lambda) \tag{3}$$

$$\mathrm{I} = \sum_{\mathrm{i}=1}^{12} i \tag{4}$$

$$i = \left(\frac{\mathrm{T}}{5}\right)^{1,514} \tag{5}$$

$$\alpha = 0,49{,}239 + 1792 \times 10^{-5}I – 771 \times 10^{-7}I^2 + 675 \times 10^{-9}I^3 \tag{6}$$

DOI: http://dx.doi.org/10.5772/intechopen.1008484

I is the annual average temperature index, i is the average monthly temperature index in °C, $F(\lambda)$, a correction factor that changes as a function of month, T is the average temperature in °C and α an index proportional to I (**Figure 5**).

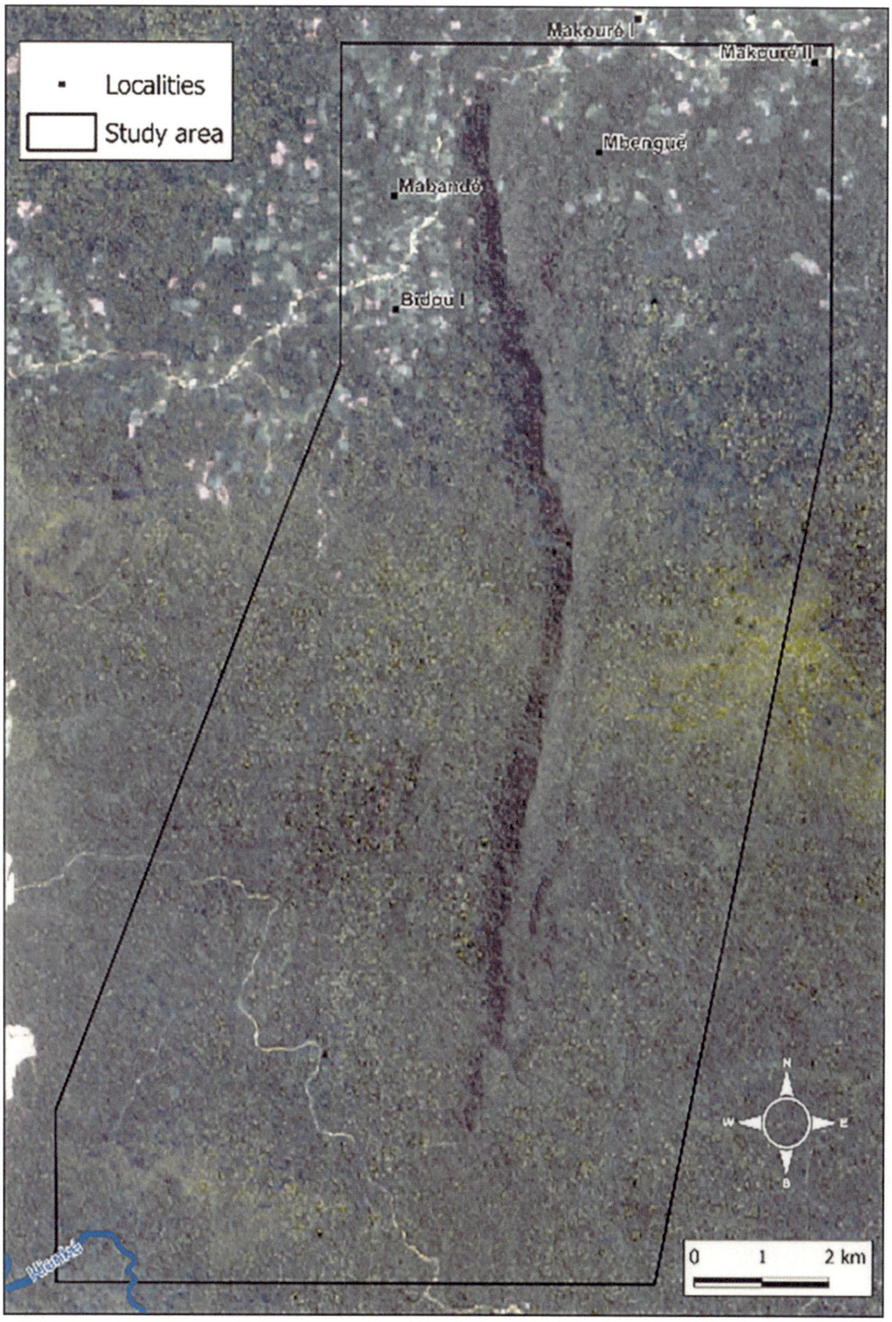

Figure 5.
Colored composite image RGB_432 of the study area.

3.2.2.4 Images filtering and lineaments extraction

As [58, 62, 64], we have used two methods for lineament extraction: the shaded relief and directional convolution screenings. From the first method, the processing parameters of the raw SRTM image are: solar incidence angle (345°), altitude (45°) and vertical exaggeration (6). In order to well visualize shaded areas, as [4, 64] shaded relief images were created for different solar trends (N45, N135, N225 and N315°) (**Figure 6**). While from the second method, the value of each pixel has been recalculated by analyzing the surrounding radiometry. The ENVI software was used to produce principal components images (PC1, PC2, PC3, PC4, PC5 and PC6) and 93% of the results is contained in PC1 after a mathematical computation from the covariance table (Eq. (1)), making it to be used for the next stages of transformation. It was then screened on PC Geomatica software from a 7x7 table following the N00, N45, N90 and N135° trends.

3.2.2.5 Mapping and validation of the lineaments map

The extracted lineaments (**Figure 7a, b**) were overlayed and combined as the result of the final lineaments map illustrating the fractures coming from the brittle tectonics. We compared the background maps (geology, road network), and the field results with all the thematic maps produced in this work to approve the final lineaments map of the study area. The same approach has been used by [4, 58, 62, 64–67].

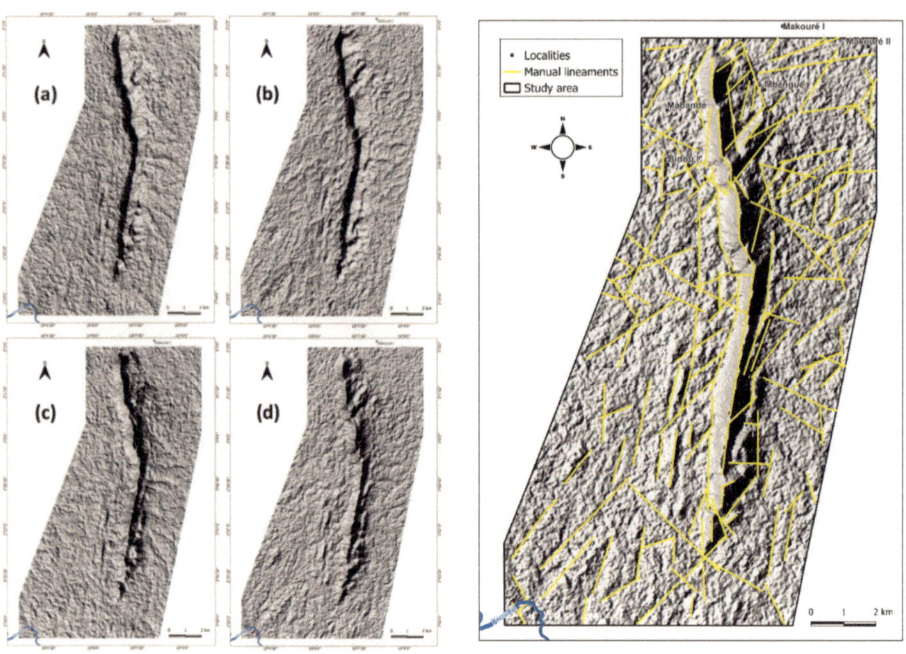

Figure 6.
DEM of the study area presented as shaded relief with different solar incidence angles: (a) 45°, (b) 135°, (c) 225°, (d) 315°, (e) manual digitized features.

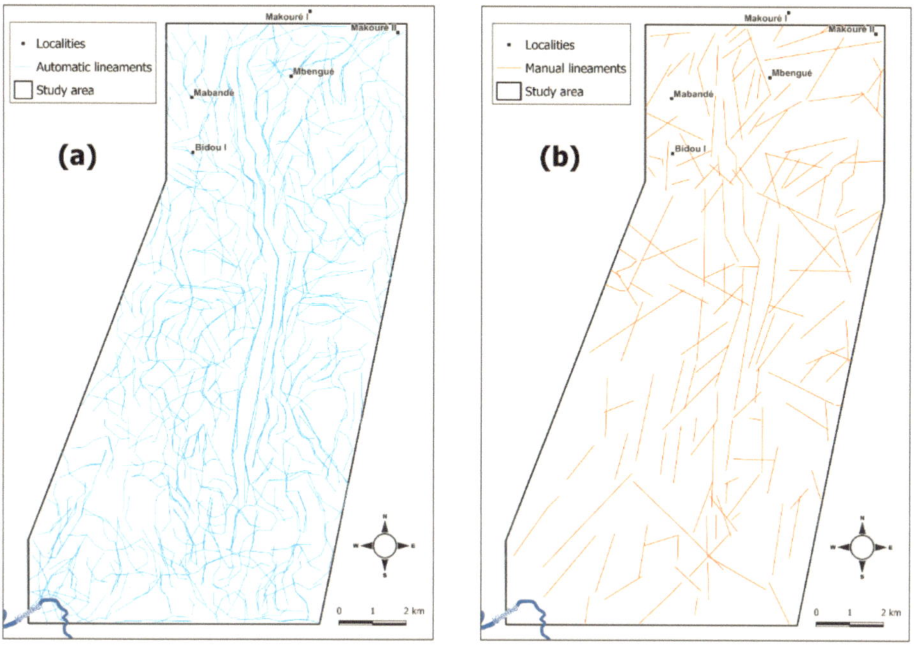

Figure 7.
(a) Automatically and (b) manually extracted lineaments.

3.2.3 Analytic hierarchy process (AHP) method and the applied notation system

The Analytic Hierarchy Process (AHP) is a robust and well-structured method for multi-criteria decision making (MCDM), based on multi-influencing factors. As one of the most widely used MCDM tools, it offers a framework for prioritizing and understanding the proportional importance of factors controlling groundwater recharge by calculating the Groundwater Recharge Potential Index (GWRPI) [11, 13, 68]. Occurrence and movement of groundwater have been governed by many variables, in this study rainfall (net recharge), geomorphology, land use/land cover (LULC), slopes, rock types, lineament density, soils and drainage density were used.

The normalized weights method modified after [2, 4, 13, 68] enabled us to calculate the weights of the various influencing factors by creating a pairwise comparison table of a x b order in which a is the number of factors and b is the number of variables. The values of the weightage were assigned based on previous works, field information and particularly the relative influence of each variable toward groundwater recharge potential. After converting all the variables in image layers, they were overlayed and compared in other to compute the Groundwater Potential Recharge Index (GWRPI). Let us be precise that, on the one hand, all these layers should have the same size and be registered in the same coordinate system of reference. On the other hand, the GWRPI is the global score calculated from the score of each factor weighted by their respective coefficients (Eq. (7)). The weights of each factor and the different classes with their ranks modified after [4] are presented in **Table 3**.

The overall score of each variable and the GWRPI are respectively computed from eqs. 7 and 9.

S. No	Factors	Weights	Classes	Ranks
1	Rainfail/Infiltration (mm/year)	5	0–50	1
			50–100	3
			100–175	6
			175–225	8
			> 225	9
2	Land cover/Land use	4	Builts-up	1
			Water bodies	9
			Vegetation	8
			Bare soils & outcrops	3
3	Soils	3	Colluvium	2
			Ferralitic	7
			Hydromorphic	4
4	Lithology	4	Schists	8
			Gneiss	6
			Amphibolites	3
			Quartzites	2
			Micaschists	7
			Duricrusts	5
			Orthogneiss	5
			BIFs	3
5	Slope (degree)	4	0–5	9
			5–9	8
			9–16	5
			16–26	3
			26–61	1
6	Drainage density map (Km/Km^2)	2	0–2	8
			2–3	7
			3–5	6
			5–6	5
			6–9	4
7	Lineament density map (Km/Km^2)	5	13–20	9
			10–13	8
			7–10	7
			4–7	6
			0–4	5

Source: modified after [4].

Table 3.
Weights, classes and ranks of each influencing parameter.

$$G = \sum_{y=1}^{a} W_y * S_{x,y} \tag{7}$$

where G is the overall score of the variables, x is the variable; y is the factor; W is variable weight; $S_{x,\ y}$ is the score of variable x for criteria y.

The AHP's results are validated if the consistency is below 10%. The value of consistency is given by the Consistency Index (CI), estimated from eq. 8 [13, 68].

$$CI = \frac{\lambda_{max} - a}{a - 1} \tag{8}$$

Where λ_{max} is the highest Eigenvalue in the comparison matrix and a is the number of factors

$$GWRPI = \frac{G}{Total\ weight} \tag{9}$$

4. Results and discussions

4.1 Analysis of the fractures network

The following figures represent the maps of all the extracted lineaments merged and the structures measured on the field. Their main directions have been determined are given by the rose diagram (**Figure 8**) and compared with the directions of the structures measured from the field (**Figure 9b**).

4.2 Spatial distribution of lineaments

A total of 2370 lineaments were collected from the visual interpretation of satellite data. The lengths varied from 60 to 4498 m, with an average of 525 ± 420 m and an overall length of 1244 km for all lineaments mapped. These linear features were classified into four groups of lengths by percentages (**Figure 10**): 225–550 m (59%); 550–1053 m (16%), 60–225 m (13%) and less than 3 km (3%). The highest cumulative length is that of the N-S oriented lineaments, followed by the NNE-SSW and NNW-SSE one, closed by the E-W oriented (**Figure 8**). Two criss-crossing networks of fractures (N-S and WNW-ESE) representing sinistral and dextral shear zones have been also exposed (**Figure 9a**) after coupled interpretation of satellite data and all field structural results.

4.3 Field works rendering

A total of 8 rock types (schists, micaschists, gneisses, amphibolites, quartzites, BIFs, orthogneisses, duricrusts) and many structures measured (metamorphic foliation, schistosity, boudins, mylonitic foliation, folds and fractures) were recognized on the field: (**Figure 11c**). These structures can be distributed into three phases of deformation (D1 to D3). The D1 phase is represented by foliations, schistosity and boudins. Meanwhile, the D2 phase is by folds, fractures and faults and the D3 by S-C fractures. Diaclases and the N-S-orientated veins of quartzites (bearing sulfides) with 80–85° dip are other field-recognized structures (**Figure 12**).

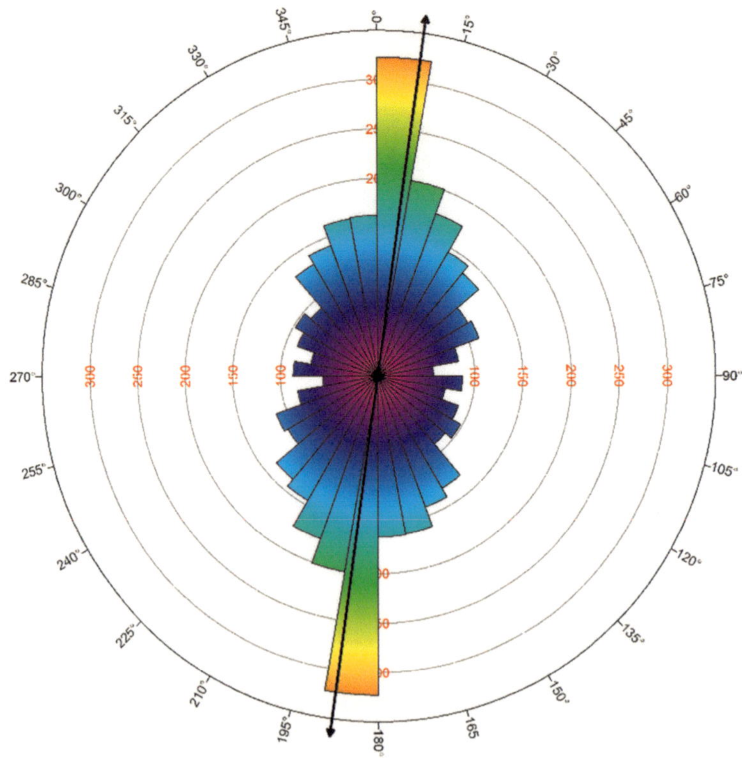

Figure 8.
Rose diagram showing lineaments according to their orientations.

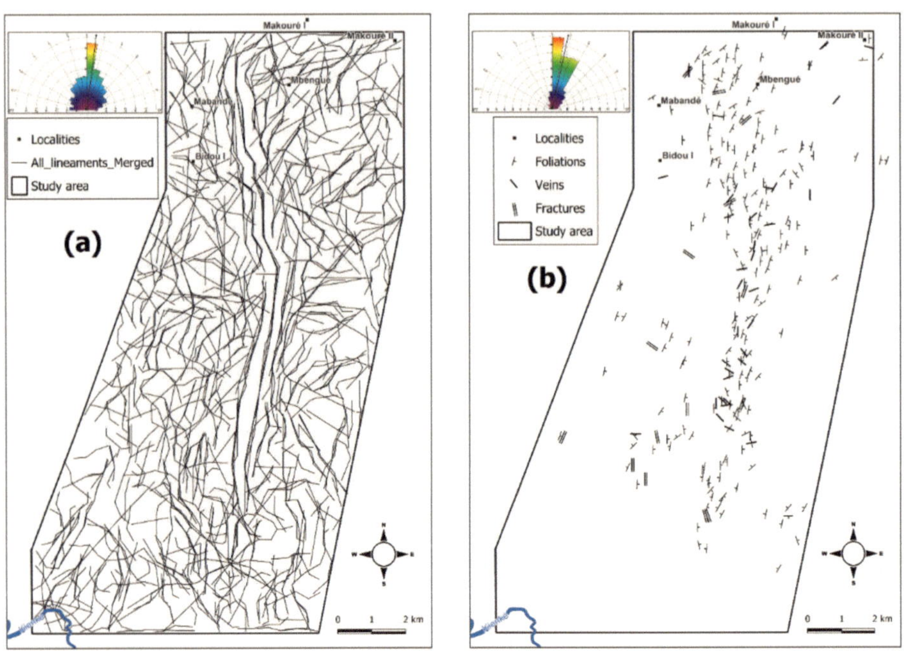

Figure 9.
(a) Map of the extracted lineaments merged; (b) map of the structures measured on the field.

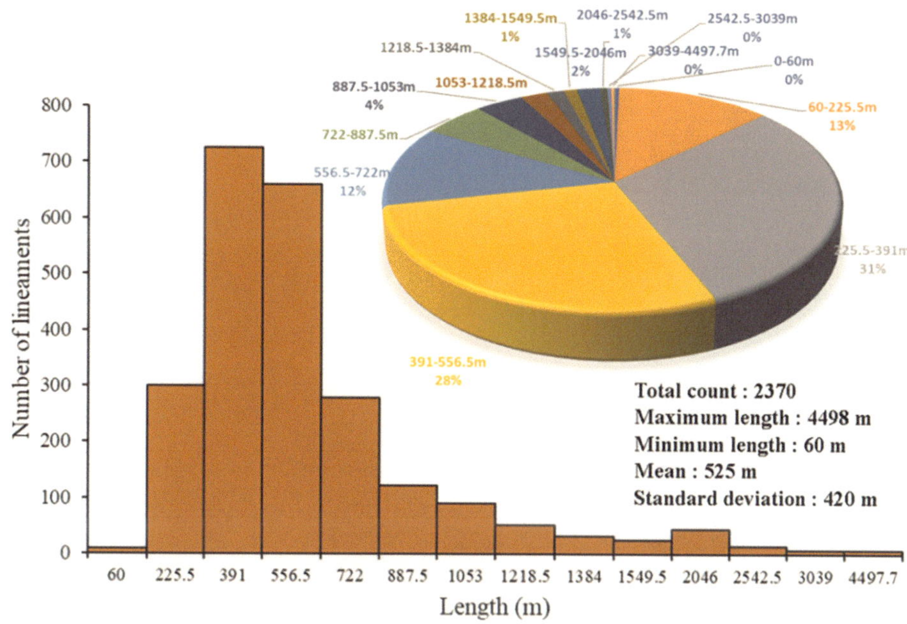

Figure 10.
Extracted lineaments distributed according to their length.

4.4 Lineament's orientation

Four families of lineaments were revealed in this work (**Figure 8**) according to the following directions: N-S, NNE-SSW, NNW-SSE and E-W with two predominant trends (N-S and NNE-SSW). Given that the lineaments orientations result from tectonic processes that produced them, the lineament's orientations in this work (between N330 and N045°) reveal those of the structures on the highly crush basement rock. Otherwise, as the groundwater movement and accumulation are governed by fractures, their orientations surely give an indication on most of the flowing directions of groundwater in the study area. As the mapped lineament's main trend is consistent with those of the structures measured on the field (**Figures 9, 13**) and the Kribi-Campo shear zone (KCSZ) oriented NNE-SSW, it is evident that the groundwater flowing directions in the study area are also between N330 and N045° with and average direction N010°E corresponding to the recharged directions of almost all aquifers in the area.

4.5 Land use/land cover (LULC)

The LU/LC distribution influences the amount of infiltration and consequently the groundwater recharge potential of the area [4, 13, 14]. **Figure 11a** displays the different classes of LU/LC in the study area, allowing the comprehension of the landscape exploitation. The land use/land cover classes of the study area agglomerate 4 macroclasses (Vegetation, Water bodies, Builts-up and Bare soils) corresponding to 10 land cover units defined as follows: gallery forests and wetlands, plantations, crops and fallow land, watercourses, bare soils and outcrops, buildings and roads. The areas

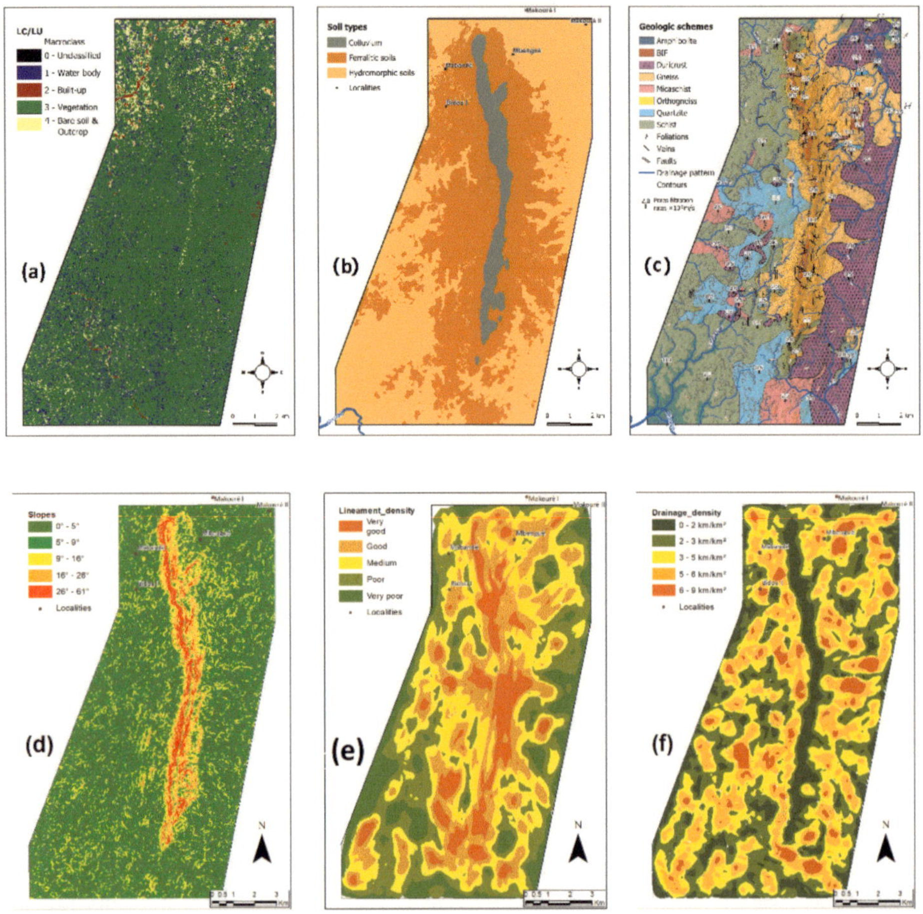

Figure 11.
(a) Land use/land cover map; (b) soils map; (c) geological map with pores filtration rates; (d) slopes map; (e) lineaments density and (f) drainage density maps of the study area.

covered by these classes, expressed as percentages of the total study area are: Vegetation: 90%, Water bodies: 5%, Bare soils: 4% and Built-up land: 1%. In southern Cameroon, forest is the main land cover type and the predominant vegetation is that of a moist evergreen and semi-deciduous equatorial forest. The southern and some eastern parts of the studied area are occupied by swamps, which reveal the permanent availability of water being part of the river network and living species of the region. Built-up areas are observed in the northern part of the area surrounded by forests and farms, showing the rate of urban development affecting the panorama.

4.6 Soils map

Many studies found soil properties as one of the most influencing factors controlling groundwater recharge [15, 16, 25–29]. Three types of soils were mapped in the study area (**Figure 11b**) by [4]: (i) the ferrallitic soils, formed on gneisses and schists. They are very porous and permeable soils with little humus. They are defined by their

DOI: http://dx.doi.org/10.5772/intechopen.1008484

Figure 12.
Field photographs from the study area: outcrops of: (a) fractured gneiss; (b) fractured schists; (c) fractured outcrop of BIF (d) diaclases with subsurface water flowing; and (e) fractured micaschist hosting a lake; (f) road trench exhibiting internal fissuring of the outcrop; (g) boudins and fractures on a weathered outcrop of schist; (h) quartz vein; (i) hand specimen of gneiss.

red clayey-sandy and very thick weathering profiles; (ii) the hydromorphic soils, coming from the continuous pressure of water table on the bedrock. Very diverse, they are found on alluvium at the foot of the hill. Accumulation of organic matter, concretion and armoring are visible in the North-Est part of the study area on the Makouré iron formations and (iii) colluvium deposited by gravity and representing the geology of the altered rocks.

4.7 Slope map

Several studies [4, 13, 14] define slopes as one of the factors that influence the flow of water from the surface into the underground. The (**Figure 11d**) of the study area

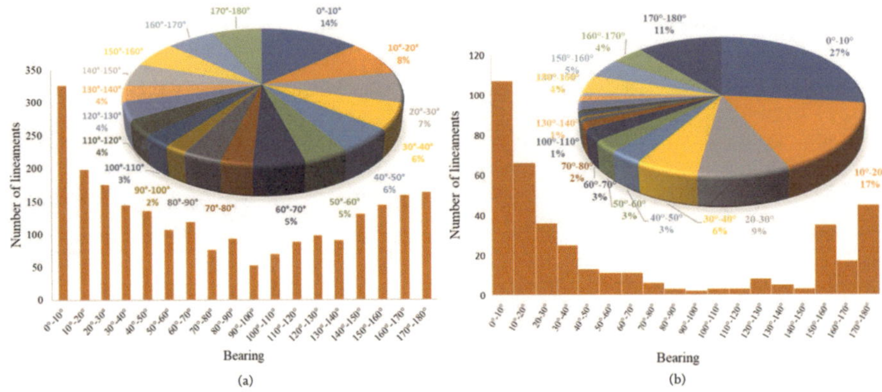

Figure 13.
(a) Extracted lineaments and (b) field structures distributed and compared according to their orientations.

indicates weak to very abrupt slope values varying between 0 and 61°, classified into five classes. The three first classes known as weak slopes (0–16°), cover about 90% of the study area, while the medium class (16–26°), covers 6% and the last class constituted by abrupt to very abrupt slopes (26–61°) which covers only 3% of the whole surface. Abrupt slopes raise the surface runoff and shorten drainage duration to which infiltration is directly connected. Meanwhile, weak slopes deplete surface runoff and allow more time for rainfall to infiltrate at depth. Hence, the area is with good GWRPZ.

4.8 Lineaments density

In hard rock areas, groundwater recharge potential is good around lineaments zones [4, 13, 26–28, 69]. As lineaments frequency expresses groundwater accumulation [69], we evaluated the number of fractures per unit area troughout the lineament density, organized in five classes (very low, low, medium, high and very high) in **Figure 11e**. This map shows that the highest lineament densities are observed along and around the Bidou I shear zone. It is why that [13] said lineaments are geological features that define the way and speed of subsurface water flow.

4.9 Lineaments intersection

The higher the density of lineaments, the easier it is for surface water to flow through these openings, and the higher the groundwater recharge potential [13]. The lineament's density map of the study area shows the frequency of joins per unit area indicating that the area is endowed with a very good groundwater recharge potential. The same results were obtained by [4, 13, 26–28, 69].

4.10 Hydrological balance of the study area

The results of the water balance presented in **Table 4** show that from 1991 to 2021 the net recharge, represents 10% of the total rainfall (2957.0 mm/year). While

	March	April	May	June	July	August	Sept.	Oct.	Nov.	Dec.	Jan.	Feb.	Total
ETP (mm)	132.3	126.9	124.3	107.5	100.8	98.1	99.3	109.2	111.8	121.1	124.3	116.6	1372.2
P (mm)	222.0	259.0	294.0	288.0	251.0	290.0	340.0	338.0	267.0	159.0	116.0	133.0	2957.0
T (°C)	26.7	26.6	26.2	25.3	24.6	24.4	24.7	25.2	25.6	26	26.2	26.5	25.7
ETR (mm)	132.3	126.9	124.3	107.5	100.8	98.1	99.3	109.2	111.8	121.1	124.3	116.6	1372.2
ETP-ETR (mm)	0.0	0.0	0.0	0.0	0.0	0.0	0.0	0.0	0.0	0.0	0.0	0.0	0.0
Ri (mm)	89.7	100.0	100.0	100.0	100.0	100.0	100.0	100.0	100.0	100.0	91.7	100.0	
ΔRi (mm)	89.7	10.3	0.0	0.0	0.0	0.0	0.0	0.0	0.0	0.0	−8.3	8.3	100.0

Table 4.
Hydrological balance of the study area.

46.4% of this total precipitation evaporated, 43.3% fed surface runoff. This, therefore, reveals the constant replacement of the water accumulated in soil. After the emptying of the water table from January to February comes the recharging in March. This is why [14] said precipitation seasonality and potential evapotranspiration, carry the most significant influence on groundwater recharge rates.

4.11 Groundwater recharge potentiality

The groundwater yields of crystalline rocks is directly linked to the degree of weathering and fracturing of the rock since their primary porosity (plutonic and metamorphic) is generally very low, and rarely exceeds 2% [70]. Thus, weathering and related fracturing tend to increase the primary porosity and permeability of affected rocks. The authors [70] speak of secondary porosity. In the works of [4], the N-S and NNE-SSW directions are defined as the main trends of structure followed by the NW-SE and NE-SW trends indicating the directions of groundwater recharge and flow movements in the area. Rocks permeability is between 0.5×10^{-9} m/s and 11.0×10^{-9} m/s with an average of 2.5×10^{-9} m/s. Furthermore, the integrated results (rock types, structural measurements, thematic maps, etc.) show that the highly fractured and weathered shales and surrounding rocks have very good to moderate groundwater recharge potential throughout the study area (**Figure 14**). Very high and high lineament density areas fit with moderate to very good GWRPZ surrounding the mountain of Bidou I. While low lineament density areas are constituted by quartzites and duricrust along the ridge and the right center part of the area fit with moderate to poor GWRPZ. Since the recharge is a function of the physical and dynamic properties of the material, a highly weathered and fractured material facilitates infiltration and thus efficient recharge and high groundwater potential [4, 70]. For [4], the more weathering and fracturing are, the more the porosity and permeability are also. Otherwise, the aquifers are recharged from the deep infiltration of part of the precipitations forming the underground provision. It is in the same order of ideas that [13, 14] said precipitation, rainfall seasonality and potential evapotranspiration, exert the most significant influence on groundwater recharge rates, with vegetation (NDVI) and topography also contributing significantly (**Figure 14**).

4.12 Validation of land use/land cover map

After the classification process, it is essential to evaluate the exactness of the land cover classification, in order to discriminate and measure cartographic lacks. To this end, an error matrix was established to compare cartographic data with ground truth data [71]. **Table 5** shows the error matrix obtained for the four (04) classes identified in the Bidou I land use/land cover classification. A total of 3348 sample units were collected. The elements of the main diagonal represent the number of correctly classified samples, while the other elements represent the classification errors. The overall accuracy calculated according to [71] formula is 92.17% (i.e. the ratio between the number of correctly classified samples (the sum of the main diagonal) and the total number of sample units). For [72] a classification is said to be good if and only if the overall accuracy is greater than 85%. If this is not the case, the classification is not good enough to represent land use and land cover, and should therefore be revised to improve accuracy. In view of the overall accuracy obtained in this classification, we

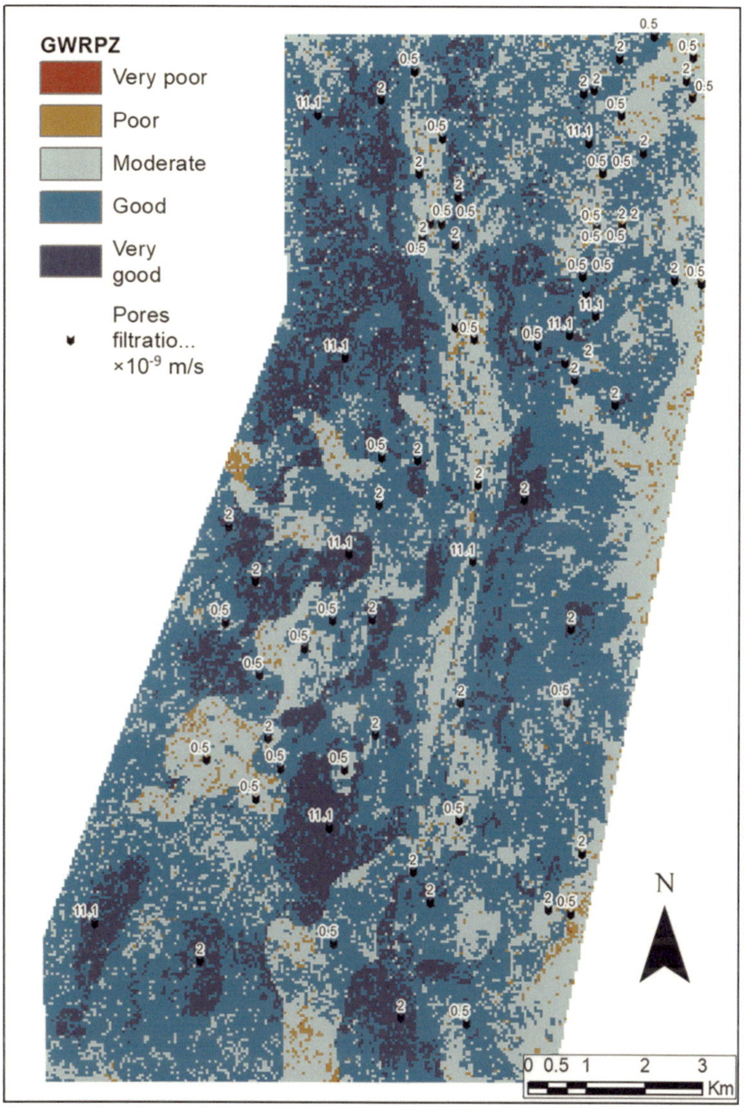

Figure 14.
Map of groundwater recharge potential zones (GWRPZ).

can say that the classification has been carried out very well and represents the LULC of the study area almost perfectly.

4.13 Validation of GWRPZ map

Contrary to [2, 13, 62], who used water level, groundwater discharge data and some empiric formulas respectively, we instead used the land cover map of the study area, the pores filtration rates, the existing lake and the all the tested boreholes to approve the resulted GWRPZ map. The overlay of all these information reveal their position on the high to moderate classes of groundwater recharge potential helping therefore to validate the correctness of the GWRPZ map of the study area.

Error matrix [pixel count]					
	> Reference				
V_Classified	1	2	3	4	Total
1	11	0	144	4	159
2	0	26	2	5	33
3	1	0	2988	24	3013
4	0	0	82	61	143
Total	12	26	3216	94	3348
AREA BASED ERROR MATRIX					
	> Reference				
V_Classified	1	2	3	4	Area
1	0.0033	0	0.0430	0.0012	143,100
2	0	0.0078	0.0006	0.0015	29,700
3	0.0003	0	0.8925	0.0072	2,711,700
4	0	0	0.0245	0.0182	128,700
Total	0.0036	0.0078	0.9606	0.0281	3,013,200
Estimated area	10,800	23,400	2,894,400	84,600	3,013,200
SE	0.0010	0.0007	0.0026	0.0025	
SE area	3025.90	2146.36	7827.13	7385.55	
95% CI area	5930.77	4206.87	15341.18	14475.67	
PA [%]	91.6667	100.0000	92.9104	64.8936	
UA [%]	6.9182	78.7879	99.1703	42.6573	
Overall accuracy [%] = 92.1744					

Table 5.
Screenshot of the accuracy assessment matrix for land use/land cover classification of Bidou I and surrounding area.

5. Conclusion

The aim of this work was to bring out groundwater recharge potential zones in Bidou I and surrounding areas, characterized by metamorphosed sedimentary rocks covered by the evergreen rainy forest of south Cameroon. To meet this goal, seven influencing parameters were merged in a GIS. Thus, we did the processing of satellite images associated with climatic data and carried out a very important fieldwork. Therefore, a total of 2370 lineaments were mapped ranging between 60 and 4498 m for a total length of 1244 km, with an average of 525 ± 420 m. The different ranges of length expressed as a percentage of the total length are: 225–550 m (59%), 550–1053 m (16%), 60–225 m (13%) and less than 3 km (3%). Eight deformed lithological types were identified on the field where structures like schistosity, metamorphic foliation, boudins, fractures and folds, forming three phases of deformation D1 to D3 were measured without forgetting diaclases and subvertical N-S orientated venous of quartzites. Four families of lineaments (N-S, NNE-SSW, NNW-SSE and E-W) were defined with two identified major fractures (N-S and WNW-ESE) cross-cutting each other. The main directions of fractures, consistent with those of the Kribi-Campo shear zone clearly demonstrate that groundwater flowing directions lie between N330° and N045°E with an average direction of N010°E, corresponding to the recharge direction of the frequently renewed aquifers. The effective infiltration of about 10% of the total rainfall in the region (2957.0 mm/year) corresponding to the stock variation (ΔRi) equivalent to the effective recharge from 1991 to 2021. The rock permeability ranged from 0.5×10^{-9} and 11.0×10^{-9} m/s with an average of 2.5×10^{-9} m/s as a function of the fractured or weathered state of the rock. The land use/land cover classes include four macroclasses (Vegetation, Water bodies, Builts-up and Bare soils) corresponding to: gallery forests and wetlands, plantations, crops and fallow land, watercourses, bare soils and outcrops, buildings and roads. The areas covered by these classes, expressed as percentages of the total study area, are: Vegetation: 90%, Water bodies: 5%, Bare soils: 4% and Built-up land: 1%. The GWRPZ map shows five classes (very poor, poor, moderate, good and very good). This map can help for a better understanding of groundwater recharge dynamics then for decision-makers it will help for drilling programs as well as sustainable groundwater protection and management in the area. Thus, the integration of Remote sensing, fieldworks merged by Analytical Hierarchy Process (AHP) approach and GIS techniques is an effective method for delineation of groundwater recharge potential zones. Being highly valuable for the society, this work by addressing critical water resource challenges it will significantly contribute to sustainable development in the region.

Acknowledgements

The authors would like to thank the editor and the anonymous reviewers for their insightful comments that helped to improve this manuscript. They also acknowledge "First published in [Sustainable Water Resources Management, 10:100, 23 pages, 2024] by Springer Nature". Reproduced with permission from Springer Nature.

Funding

No funds, grants, or other support was received.

Conflicts of interest

We declare no conflict of interest regarding this chapter.

Author details

Olivier Njikeu[1*], Roger Feumba[2] and Benjamin Ngounou Ngatcha[1]

1 University of Ngaoundéré, Ngaoundéré, Cameroon

2 University of Yaoundé I, Yaoundé, Cameroon

*Address all correspondence to: o.njikeu@yahoo.fr

References

[1] Feumba R. Hydrogéologie et évaluation de la vulnérabilité des nappes dans le bassin versant de Besseke (Douala, Cameroun) [thesis]. Cameroun: Université de Yaoundé I; 2015. 255p

[2] Gupta DS, Biswas A, Ghosh P, Rawat U, Tripathi S. Delineation of groundwater potential zones, groundwater estimation and recharge potentials from Mahoba district of Uttar Pradesh, India. FEBS Letters. 2021. 25p. DOI: 10.1007/s13762-021-03795-0

[3] Kesarwani M, Neeti N, Chowdary VM. Evaluation of different gridded precipitation products for drought monitoring: A case study of Central India. Theoretical and Applied Climatology. 2023;**151**:817-841

[4] Njikeu O, Feumba R, Kemgang Dongmo T, Deschamps P, Ngounou Ngatcha B. Groundwater potential zones mapping by combining remote sensing, modified DRASTIC method and field data for Bidou I, South Cameroon. Sustainable Water Resources Management. 2024;**10**:100. DOI: 10.1007/s40899-024-01031-4

[5] Ngounou NB. Hydrogéologie d'aquifères complexes en zone semi-aride: les aquifères quaternaires du Grand Yaéré (Nord Cameroun) Géologie appliquée [thesis]. Français: Université Joseph-Fourier-Grenoble I; 1993. fftel-00783502, 352p

[6] Ngounou Ngatcha B, Mudry J, Aranyossy JE, Naah E, Sarrot RJ. Apport de la géologie, de l'hydrogéologie et des isotopes de l'environnement à la connaissance des « nappes en creux » du grand Yaéré (Nord Cameroun). Revue des sciences de l'eau 20. 2007a;**1**:29-43. DOI: 10.7202/014905ar

[7] Chinnasamy P, Maheshwari B, Prathapar S. Understanding groundwater storage changes and recharge in Rajasthan, India through remote sensing. Water. 2015;7(10): 5547-5565. DOI: 10.3390/w7105547

[8] Shah D, Mishra V. Integrated drought index (IDI) for drought monitoring and assessment in India. Water Resources Research. 2020;**56**:e2019WR026284. DOI: 10.1029/2019WR026284

[9] Djeuda Tchapnga HB, Tanawa E, Ngnikam E. L'eau au Cameroun. Tome 1: Approvisionnement en eau potable. In: Laboratoire Environnement et Sciences de l'Eau (LESEAU), École Nationale Supérieure Polytechnique de Yaoundé. In: Presses Universitaire de Yaoundé. Yaoundé: Wiley; 2001. 359p

[10] Patel NR, Pokhriyal S, Singh RP. Advancements in remote sensing based crop yield modelling in India. Journal of Agrometeorology. 2023; **25**:293-299

[11] Elsebaie IH, Kawara AQ. Modeling groundwater recharge potential zones in the Wadi Yalamlam, Saudi Arabia. Frontiers in Water. 2024;**6**:1387741. DOI: 10.3389/frwa.2024.1387741

[12] Chowdhury A, Jha MK, Chowdary VM. Delineation of groundwater recharge zones and identification of artificial recharge sites in West Medinipur district, West Bengal, using RS, GIS and MCDM techniques. Environmental Engineering Science. 2010;**59**:1209-1222. DOI: 10.1007/s12665-009-0110-9

[13] Thakur J, Singh SK, Ekanthalu VS. Integrating remote sensing, geographic information systems and global

positioning system techniques with hydrological modeling. Applied Water Science. 2019;7:1595-1608. DOI: 10.1007/s13201-016-0384-5

[14] Vinay A, Mavidnam SR. Groundwater recharge potential index and artificial groundwater recharge in the alluvial soils of the middle Ganga Basin. Case Study, Discover Applied Sciences. 2024;**6**:367. DOI: 10.1007/s42452-024-05851-z

[15] Lee S, Irvine DJ, Duvert C, Rau GC, Cartwright I. A high-resolution map of diffuse groundwater recharge rates for Australia. Hydrology and Earth System Sciences. 2024;**28**:1771-1790. DOI: 10.5194/hess-28-1771-2024

[16] Njiké Ngaha PR. Contribution à l'étude géologique, stratigraphique et structurale de la bordure du bassin Atlantique du Cameroun [thesis]. Cameroun: Université de Yaoundé I; 1984. 133p

[17] Sighomnou D. Analyse et redéfinition des régimes climatiques et hydrologiques du Cameroun : perspectives d'évolution des ressources en eau [thesis]. Cameroun: Université de Yaoundé I; 2004. 291p

[18] Anaba Onana AB. Caractérisation hydrodynamique et analyse de la productivité des aquifères de fissures et de fractures du socle cristallin : cas de la Région du centre au Cameroun [thesis]. Cameroun: Université de Yaoundé I; 2017. 227p

[19] Mafany GT, Fantong WY. Groundwater quality in Cameroon and its vulnerability to pollution. In: Xu Y, Usher B, editors. Groundwater quality in Africa. Taylor and Francis/Balkema, Rotterdam: Wiley; 2006. pp. 47-55

[20] MacDonald AM, Bonsor HC, Dochartaigh BEO, Taylor RG. Quantitative maps of groundwater resources in Africa. Environmental Research Letters Willey. 2012;7:024009

[21] Ketcheumen TB. Etude hydrogéologique du Grand Yaéré (Extrême-Nord du Cameroun) : Synthèse hydrogéologique et étude de la recharge par les isotopes de l'environnement. Géologie appliquée [thesis]. Sénégal: Université Cheikh Anta-Diop de Dakar; 1992. 216p

[22] Njitchoua R, Dever L, Fontes JC, Naah E. Geochemistry, origin and recharge mechanisms of groundwaters from the Garoua Sandstone aquifer, northern Cameroon. Journal of Hydrology. 1997;**190**:123-140

[23] Kemgang Dongmo T, Boucher M, Ngounou Ngatcha B, Favreau G, Mvondo VYE, Ibrahim M, et al. Magnetic resonance soundings for characterizing the quaternary aquifer in the lakechad basin– a case history from Cameroon and Niger. Water Resources. 2015;**25**:69-91

[24] Mvondo V, Kemgang Dongmo T, Favreau G, Ngounou Ngatcha B. Use of Geological and Hydrogeochemical Data to Investigate Depressed Aquifers in the Southern Part of Lake Chad Basin (LCB), Cameroon. American Journal of Water Resources. 2018;**6**(4):169-175. DOI: 10.12691/ajwr-6-4-4

[25] Kemgang, Dongmo T, Boucher M, Mvondo VYE, Favreau G, Ngounou Ngatcha B, Yalo N, et al. Contribution of time domain electromagnetic and magnetic resonance soundings to groundwater assessment at the margin of lake Chad basin, Cameroon. Journal of Applied Geophysics. 2019. 23p. DOI: 10.1016/j.jappgeo.2019.103840

[26] Ewodo Mboudou GA, Ombolo Fouepe Takounjou A, Bon AF,

Ekodeck GE. Étude des Paramètres hydrauliques des aquifères de sub-surface du bassin versant de la Mingosso, région de Yaoundé. Révue-Conseil Africain et Malgache pour l'Enseignement Supérieur (CAMES). 2012;**13**:123-127

[27] Bon AF, NdamNgoupayou JR, Ewodo Mboudou G, Ekodeck GE. Caractérisation hydrogéologique des aquifères de socle altéré et fissuré du bassin versant de l'Olézoa à Yaoundé, Cameroun. Revue des sciences de l'eau. 2016;**29**:149-166. DOI: 10.7202/1036545ar

[28] Anaba, Onana AB, Njikeu O, Onana Onana RM. Finding parameter that provide higher boreholes fow rate: using ANOVA test, Cameroon. Sustainable Water Resources Management. 2020;**6**: 94. DOI: 10.1007/s40899-020-00451-2

[29] Feumba R, Kemgang Dongmo T, Tabué Youmbi JG, Mvondo VYE, Tchaptchet Tchato DI, Ngounou NB. Assessment of hydraulic parameters and protection zones of catchment aquifers for water supply network in Ndé Division, West-Cameroon (Central Africa). Journal of Water Resource and Protection. 2021;**13**:478-497. DOI: 10.4236/jwarp.2021.137028

[30] Mondiale B. Evaluation hydrologique de l'Afrique Sub-Saharienne, Pays de l'Afrique de l'Ouest. Rapport de pays Cameroun réalisé Mott MacDonald International, BCEOM: SOGREAH and ORSTOM; 1992

[31] SOGREAH. Inventaire de sites de barrages dans les Mont Mandara - Données de base, Carte 1/50.000 Hydrogéologie du périmètre Tsanaga. Montpellier, France: ORSTOM; 1976.

[32] CIEH-BRGM. Carte de planification des ressources en eau du Cameroun – FAC 2 feuilles au 1/1000.000. Orleans Cedex, France: BRGM; 1979. 72p

[33] Wakuti. Etude hydrogéologique du bassin crétacé de la Benoué, FED/ET - 824. Siegen RFA: Rapport Wakuti; 1968

[34] Savadogo AN. Géologie et hydrologie du socle cristallin de Haute-Volta : étude régionale du bassin versant de la Sissili [thesis]. France: Université de Grenoble I; 1984. 351p

[35] Djeuda Tchapnga HB. Géologie et hydrogéologie d'un secteur de la zone mobile d'Afrique centrale : région de poli, nord – Cameroun [thesis]. France: Univ. sci. tech. et méd. de Grenoble I; 1987. 333p

[36] CIEH. Méthode d'étude et recherche de l'eau souterraine des roches cristallines de l'Afrique de l'ouest. Goo hydraulique. 1981;**38**:100

[37] Nzenti JP, Barbey P, Bertrand JML, Macaudiere J. La chaîne panafricaine au Cameroun: cherchons suture et modèle. In: S.G.F. édit., 15e réunion des Sciences de la Terre. Nancy, France: Science and Education Publishing; 1994. 99p

[38] Nzenti JP, Barbey P, Tchoua FM. Mint: Evolution crustale au Cameroun: éléments pour un modèle géodynamique de l'orogenèse néoproterozoïque. In: Dans: Vicat JP, Bilong P, editors. In géologie de l'environnement au Cameroun, Collection Geocam 2: Book In: Presse de l'Université de Yaoundé I, Yaoundé-Cameroun; 1999. pp. 397-407

[39] Maurizot P, Abessolo A, Feybesse JL, Johan LP. Etude de prospection minière du Sud-Ouest Cameroun. Synthèse des travaux de 1978 à 1985. Rapport De BRGM. 1986;**85**:274

[40] Tchameni R. Géochimie et géochronologie des formations de

l'Archéen et de Paléoprotérozoique du Sud Cameroun (Groupe du Ntem, Craton du Congo) [thesis]. France: Université d'Orléans; 1997. 395p

[41] Tchameni R, Mezger K, Nsifa NE, Pouclet A. Crustal origin of early proterozoic syenites in the Congo Craton (Ntem Complex), South Cameroon. Lithos. 2001;**57**(1):23-42

[42] Lerouge C, Cocherie A, Toteu SF, Penaye J, Milési JP, Tchameni R, et al. Shrimp U-Pb zircon age evidence for Paleoproterozoic sedimentation and 2.05 Ga syntectonic plutonism in the Nyong Group, South-Western Cameroon: consequences for the Eburnean-Transamazonian belt of NE Brazil and Central Africa. Journal of African Earth Sciences. 2006;**44**:413-427

[43] Shang CK, Liégeois JP, Satir M, Frisch W, Nsifa EN. Late Archaean high-K granite geochronology of the northern metacratonic margin of the Archaean Congo craton, Southern Cameroon: evidence for Pb-loss due to non-metamorphic causes. Gondwana Research. 2010;**18**(2–3):337-355

[44] Toteu SF, Van Schmus WR, Pénaye J, Nyobe JB. U-Pb and Sm–Nd evidence for Eburnean and Pan-African high-grade metamorphism in cratonic rocks of southern Cameroon. Precambrian Research. 1994;**67**:321-347

[45] Thiart C, de Wit MJ. Fingerprinting the metal endowment of early continental crust to test for secular changes in global mineralization. Geological Society of America. Memory of the Society. 2006;**198**. DOI: 10.1130/2006.1198(03)

[46] Feybesse JL, Johan V, Triboulet C, Guerrot C, Mayaga-Mikolo F, Bouchot V, et al. The West Central African belt: a model of 2.5–2.0 Ga accretion and two-phase orogenic evolution. Precambrian Research. 1998; **87**:161-216

[47] Penaye J, Toteu SF, Tchameni R, VanSchmus WR, Tchakounté J, Ganwa A, et al. The 2.1 Ga west Central African Belt in Cameroon: extension and evolution. Journal of African Earth Sciences. 2004;**39**:159-164

[48] Kankeu B, Greiling RO, Nzenti JP, Ganno S, Danguene PYE, Bassahak J, Hell JV. Mint: Contrasting Pan-African structural styles at the NW margin of the Congo Shield in Cameroon. Journal of African Earth Sciences. 2018; **146**;28-47. DOI: 10.1016/j.jafrearsci.2017.06.002

[49] Bouyo Houketchang M, Penaye J, Mouri H, Toteu SF. Eclogite facies metabasites from the Paleoproterozoic Nyong Group, SW Cameroon: mineralogical evidence and implications for a high- pressure metamorphism related to a subduction zone at the NW margin of the Archaean Congo craton. Journal of African Earth Sciences. 2019; **149**:215-234

[50] Nga Essomba Tchoungui PE, Ganno S, Tanko Njiosseu EL, Ndema Mbongue JL, Kamguia Woguia B, Takodjou Wambo JD, et al. Geochemical constraints on the origin and tectonic setting of the serpentinized peridotites from the Paleoproterozoic Nyong series. Eseka area, SW Cameroon. Acta Geochimica. 2020; **39**(3):404-422. DOI: 10.1007/s11631-019-00368-4

[51] Pouclet A, Tchameni R, Mezger K, Vidal M, Nsifa EN, Shang CK, et al. Archaean crustal accretion at the northern border of the Congo craton (South Cameroon), The charnockite-TTG link. Bulletin of the Geological Society of France. 2007;**178**:331-342

[52] Ganno S, Tsozué D, Kouankap Nono GD, Tchouatcha MS, Ngnotué T, Takam GR, et al. Geochemical constraints on the origin of banded iron formation hosted iron ore from the Archaean Ntem Complex (Congo Craton) in the Meyomessi Area, Southern Cameroon. Resource Geology. 2018;**68**(3):287-302

[53] Owona S, Ratschbacher L, Nsangou Ngapna M, Gulzar AM, Mvondo Ondoa J, Ekodeck GE. How diverse is the source? Age, provenance, reworking, and overprint of Precambrian meta-sedimentary rocks of West Gondwana, Cameroon, from zircon U-Pb geochronology. Precambrian Research. 2021;**359**:106220

[54] Kamguia Woguia B, Kouankap Nono GD, Azinwi Tamfuh P, Embui Fonabe V, NforbaTamnta M, Nzenti JP. Identifying multi-metal prospect using regional soil and stream sediment geochemical data in Bidou, Nyong Series, North West of Congo Craton. Arabian Journal of Geosciences. 2021;**14**:218. DOI: 10.1007/s12517-021-06579-5

[55] Kamguia Woguia B, Kouankap Nono GD, Nga Essomba Tsoungui PE, Tanko Njiosseu EL, Ayonta Kenne P, Nzenti JP. Geochemistry and U–Pb zircon age of the Paleoproterozoic metasedimentary rocks from the Bidou I, Nyong Series, Cameroon: Implications for provenance and tectonic setting. Arabian Journal of Geosciences. 2022;**15**: 154. DOI: 10.1007/s12517-022-09476-7

[56] Sepwouo Sawouo ADS, Kouankap Nono GD, Mvodo H, Nyemb Sayom P, Kamguia Woguia B, Nzenti JP. Petrography and geophysical characterization of the newly discovered iron ore deposit in Makoure (Bidou-East), Nyong group South Cameroon. Results in Geophysical Sciences. 2022;**12**: 100050

[57] Sepwouo, Sawouo ADS, Kouankap Nono GD, Njiosseu Tanko EL, Njikeu O, Ngouem PA, Kamguia Kamani MS, et al. Alluvial gold particles microchemistry and induced polarization geophysical surveys in Makouré area, Nyong group, Cameroon: Constrain of orogenic gold prospectivity. Arabian Journal of Geosciences. 2023a;**16**:359. DOI: 10.1007/s12517-023-11452-8

[58] Aminov J, Chen X, Anming B, Aminov J, Mamadjanov Y, Aminov J, et al. Comparison of multi-resolution optical Landsat-8, Sentinel-2 and Radar Sentinel-1 data for automatic lineament extraction: A case study of Alichur Area, SE Pamir. Remote Sensing. 2019;**11**:778. DOI: 10.3390/rs11070778

[59] Veeraraghavan V. A Quantitative Analysis of Pansharpened Images [Master Degree]. Starkville: Mississippi State University; 2004. 94p

[60] Alcaras E, Parente C, Vallario A. Automation of pan-sharpening methods for Pléiades images using GIS basic functions. Remote Sensing. 2021;**13**:1550. DOI: 10.3390/rs13081550

[61] Adiri Z, elHarti A, Jellouli A, Lhissou R, Maacha L, Azmi M, et al. Comparison of landsat-8, ASTER and Sentinel 1 satellite remote sensing data in automatic lineaments extraction: A case study of Sidi Flah-Bouskour Inlier, Moroccan Anti Atlas. Advances in Space Research. 2017;**60**:2355-2367. DOI: 10.1016/j.asr.2017.09.006

[62] Rami D, Mohamed C, El HadjYoucef B. Structural lineament mapping using remote sensing in the aures massif (NE Algeria). Global Journal Of Geological Sciences. 2023;**21**: 129-142

[63] Thornthwaite CW. An approach toward a rational classification of climate. Geography Review. 1948;**38**: 55-94. DOI: 10.2307/210739

[64] Mouncherou OF, Njikeu O, Kamtchueng B, Kpoumié A, Zakari Mfonka Z, Mfochive OF, et al. Use of landsat 8 OLI images to assess groundwater potential areas in the Bamun Plateau: Cameroon Volcanic Line (CVL). Journal of Water Resource and Protection. 2020;**12**:558-576. SSN Online: 1945-3108

[65] Eskicioglu A, Fisher P. Image quality measures and their performance. IEEE Transactions on Communications. 1995; **43**:2959-2965

[66] Prasad RK, Mondal NC, Pallavi B, Nandakumar MV, Singh VS. Deciphering potential groundwater zone in hard rock through the application of GIS. Environmental Geology. 2008;**55**: 467-475. DOI: 10.1007/s00254-007-0992-3

[67] Raouf A, Yue J, Njeudjang K, Soh Tamehe L, Onibudo OO. Remote sensing and geophysical methods for delineating groundwater resources in schists hard rocks of the Adamawa Yadé Domain, Cameroon. The Egyptian Journal of Remote Sensing and Space Sciences. 2023;**26**:217-230

[68] Rajesh SP, Umakant R, DevSen G, Arkoprovo B, Shashikant T, Parthapratim G. Assessment of groundwater potential zones using multi-criteria evaluation technique of Paisuni River Basin from the combined state of Uttar Pradesh and Madhya Pradesh, India. Environmental Earth Sciences. 2020;**79**:340. DOI: 10.1007/s12665-020-09091-3

[69] Srivastava PK, Bhattacharya AK. Groundwater assessment through an integrated approach using remote sensing, GIS and resistivity techniques: A case study from a hard rock terrain. International Journal of Remote Sensing. 2006;**27**(20):4599-4620

[70] Banton O, Bangoy LM. Hydrogéologie: Multiscience environnementale des eaux souterraines. Book In: Presse de l'Université du Québec/AUPELF-UREF, Canada; 1997. 460p

[71] Congalton R, Green K. Assessing the Accuracy of Remotely Sensed Data: Principles and Practices. Boca Raton, FL: CRC Press; 2009

[72] Congedo C. Semi-automatic classification plugin: A python tool for the download and processing of remote sensing images in Qgis. Journal of Open Sources Software. 2021;**6**(64):3172. DOI: 10.21105/joss.03172

Section 2

Advanced Hydro-Geophysical Study

Chapter 4

Hydrogeophysical Investigations for Groundwater Potentiality in Arid and Semiarid Zones: A Case Study in the Upper River Atbara Watershed, Eastern Sudan

Khalid Nayl, Abdalla Elsheikh, Adil Elkrail and Abobaker Elbahari

Abstract

This chapter is focusing on the application of hydro-geophysical methods to investigate the groundwater occurrences in the area between Upper Atbara, Setit, and Khashm El Girba dams, Eastern Sudan. Vertical Electrical Sounding Techniques were applied using Schlumberger Array in which the study area was covered by (177) conducted points of Vertical Electrical Soundings (VES). The VES methods are used to obtain detailed information about the depth to the basement rocks, find out the resistivity of the different subsequent layers, identify aquifers thickness/boundaries, and select suitable sites for drilling successful boreholes. Eight geo-electrical sections were constructed in the study area in E-W and in N-S directions. The geophysical investigations confirmed that the study area consisted of seven geological units: superficial deposits, unconsolidated Karab formation, Rivers Atbara and Setit sediments, Cenozoic basalts, cretaceous sedimentary rocks, and the basement complex. The study area was divided into upstream, middle, and downstream areas. The variations in the apparent resistivity values indicate various formations and layers. The study results revealed that the depth to the basement rocks occurred between 40 and 200 m below surface. The aquifers thickness varies between 20 and 80 m. The main water-bearing formations in the study area are the Cretaceous Sandstone, River Atbara sediments, and the fractured basalt near rivers.

Keywords: groundwater, geophysics, resistivity method, river Atbara, Sudan

1. Introduction

The geophysical techniques are well-known in geological exploration fields as potential or direct applied methods such as gravity, resistivity, seismic, and act.

The geophysical methods have difference response to physical properties (anomalies) within the earth's crust. Geophysical methods can help in solving the geological problem through the detection of occurrence, thickness, and depth of the subsurface layers [1].

Geophysical methods can be very effective in groundwater resources exploration; hence, several geophysical techniques are commonly applied in groundwater investigations. The Electromagnetic and Electrical Resistivity methods are the most common geophysical methods used in groundwater exploration.

The theory of the electrical method is well explained in standard texts such as references [2–5].Resistivity is the most appropriate geophysical method used in groundwater exploration. This technique measures the resistance of the geological formations. However, resistivity generally has lower resolution than TDEM

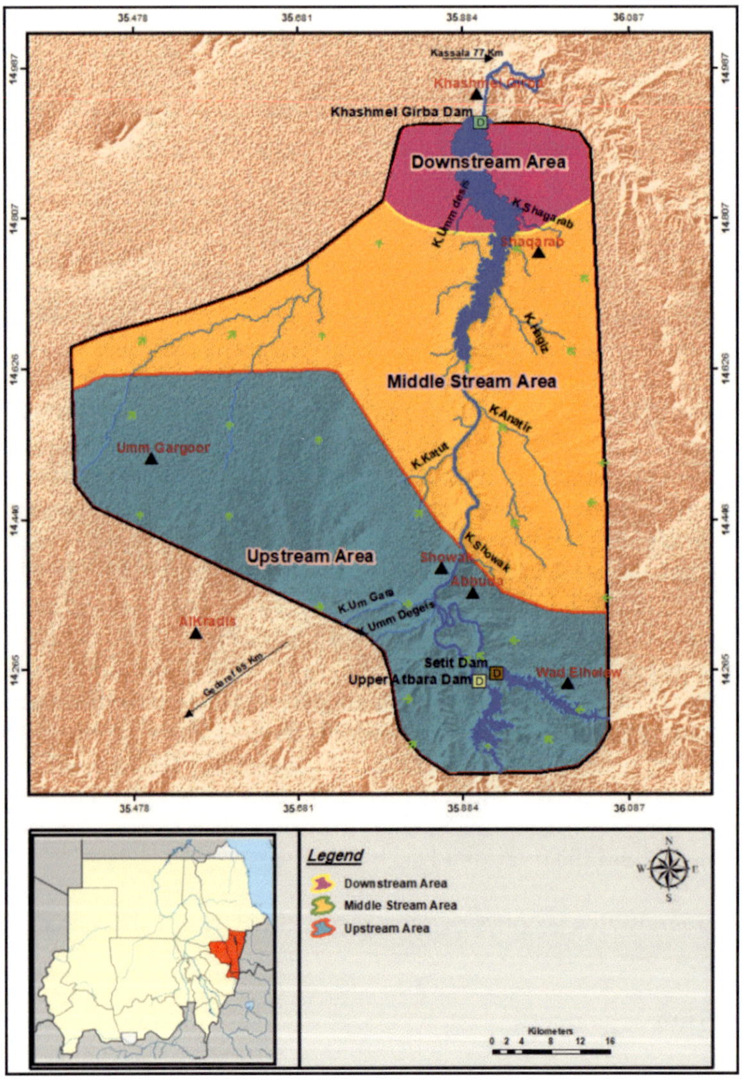

Figure 1.
Location of the study area sectors.

techniques, in both vertical and lateral extensions, but resistivity is a comparatively available and cheap method.

Resistivity can be an effective shallow sounding method, although the electrical method are not well located to civil area due to interference from structures, fences, and utilities [6].

The eastern part of the study area is situated in Kassala state, while the western part located in Gedaref state, eastern Sudan. The study area is enclosed by latitudes 14.134170°N to 14.961004° N and longitudes 35.536336° E to 35.962208° E. The study area is conforming irregular shape of about 3000 Km^2, (**Figure 1**).

The main objective of this chapter is focusing on the application of hydro-geophysical methods to investigate the groundwater occurrences in the area between Upper Atbara, Setit, and Khashm El Girba dams in eastern Sudan.

2. Electrical resistivity methods

The Electrical Resistivity method (ER) is the well-known geophysical method applied in groundwater exploration field that is used for deep and shallow subsurface groundwater exploration. The Vertical Electrical Sounding measurements determine the thickness and resistivity values of different subsurface layers and aquifer zones [7].

2.1 Basic concepts

The electrical resistivity method had been used in the current investigation in which Vertical Electrical Sounding (VES) techniques are thought to be as the most significant in groundwater exploration to detect the vertical variations of subsurface formations.

The nondestructive subsurface mapping technique is used for detecting subsurface phenomena such as underground water, cracks, buried channels, faults, fractured, and weak weathered zones. The basis of the method employs measurements of an electrical potential associated with subsurface electrical current flow generated by a direct or slightly alternating Current. The electric current is sent to the subsurface units through two current electrodes and measurement of the potential at the surface through two potential electrodes [8].

The VES technique is applied to determine the vertical variations in electrical resistivity values, whereas profiling technique is used to detect lateral changes in electrical resistivity. The lateral variations are attributed to the variations in lithology, degree of weathering, fracturing, and their water content [9]. The limitation of the resistivity sounding method is due to horizontal (or lateral) change in the subsurface resistivity [10].

The most common arrays used in resistivity surveys are Schlumberger, Wenner, Dipole-Dipole, and Lee-Partition Spread that depends on the survey portrayal [11]. The choice of the "best" array for field survey depends on the formation typeset and on the sensitivity of the resistivity meter [10].

Ohm's Law defines the association between the conducted current through a wire and the potential voltage requisite to drive the current.

$$V = IR \tag{1}$$

Where V is the voltage (volts), I is the current (amperes), and R is the resistance (ohms) [12].

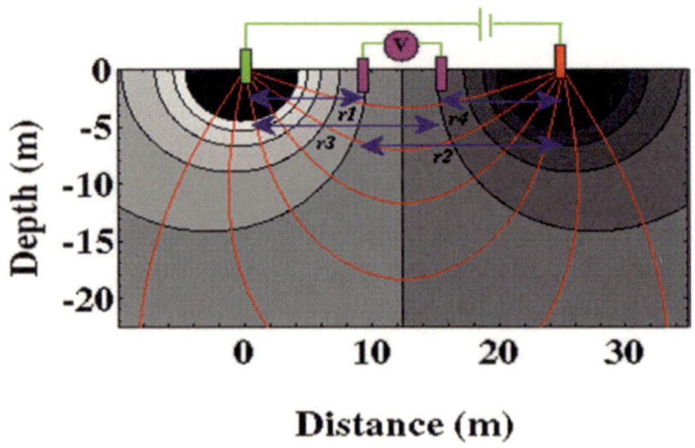

Figure 2.
The distribution of electrodes according to Schlumberger configuration.

The distances between the four electrodes is given by r1, r2, r3, and r4, as shown in **Figure 2**; then, the resistivity is calculated according to the equation below:

$$p_a = \frac{2\pi\Delta V}{i}\left[\frac{1}{\frac{1}{r1}-\frac{1}{r2}-\frac{1}{r3}-\frac{1}{r4}}\right] \tag{2}$$

The measured resistivity is called apparent resistivity, and its value is the resistivity of a homogenous earth material, [13]. The relationship between true and apparent resistivity is complicated. The subsurface resistivity is affected by many factors including rock types, properties and their contents, water composition, formation homogeneity, and other geological and topographical rudiments [14]. The ground resistivity is related to various lithological parameters such as porosity, contents, and the degree of saturation [15].

The geophysical methods are always using to image the subsurface geology and to reduce the high cost of drilling [16]. Generally, the crystalline rocks (metamorphic and igneous) have high resistivity values, where the degree of fracturing and its water contents will reduce their resistivity. The sedimentary rocks are usually porous and higher water content discloses relatively low resistivity values. Clayey soils normally have a relative lower resistivity values compared with sandy soil. The resistivity of groundwater depends on the concentration of dissolved salts, which in general vary from 20 to 100 Ωm (Appendix A).

The clue of the application of the electrical sounding method in this investigation is to identify resistivity zones that are related to adequate groundwater occurrences since the VES data reflect the expectation of groundwater occurrence within different geological formations.

2.2 Data acquisition

Schlumberger Array was adopted for the geophysical survey to attain the maximum possible depths, with a maximum possible half separation (AB/2 = 800–1000 m). In this array, the four electrodes are situated in straight line directly on the surface in A M N B order (**Figure 3**).

The study area is covered by (177) of (VES) points distributed to cover the study area (**Figure 4**). The coordinates of all VES points have been fixed by GPS device

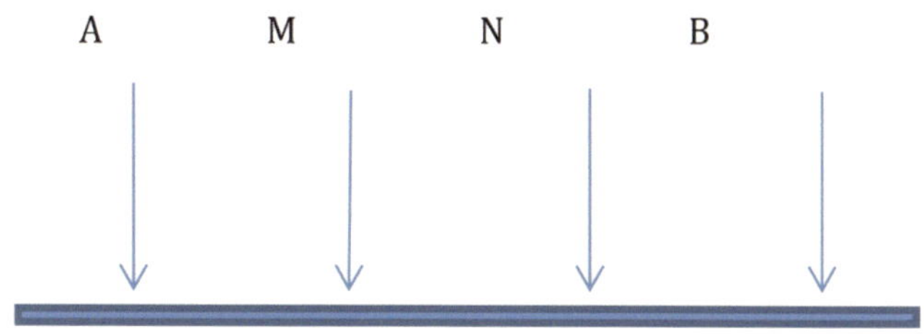

Figure 3.
Schlumberger arrays.

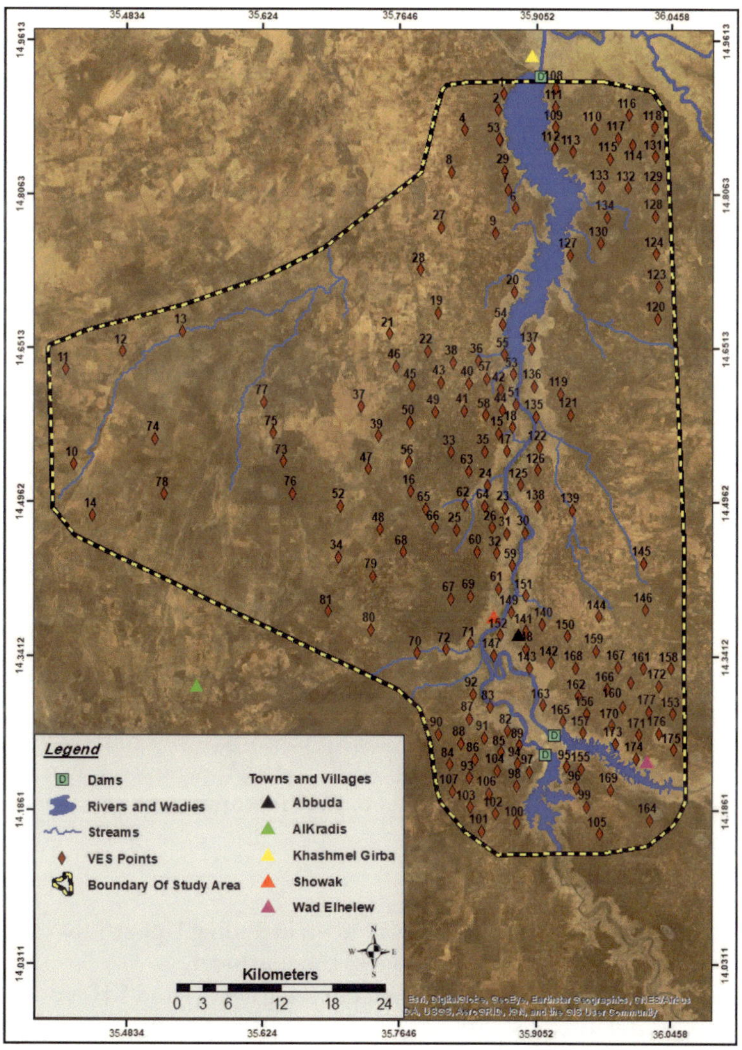

Figure 4.
Location of the vertical electrical sounding points.

(Appendix B). The conductance of VES points was distributed to fill the gabs in areas of no boreholes. During this investigation, an ABEM SAS-1000 Terrameter Instrument (Sweden) was used for VES measurements.

2.3 Data interpretation

The shapes of the plotted data curves give qualitative analysis, where measuring the layers resistivity values and depths gives quantitative analysis. The free source geophysical analytical software (IPI_{2WIN}) from Moscow University was used for data interpretation.

2.3.1 Qualitative interpretation

The geological knowledge of the study area is the chief to guide the geophysical data interpretation. The qualitative interpretation concerned the curve shapes on the apparent resistivity of various formations.

The field curves generally represent HA, AH, HKH, and KH curve types, demonstrating 6–7 layers portray thick layers of relatively low resistivity values occurred in the upstream area. In the downstream area, A-type curves are common, indicating 3–4 layers and the existence of basaltic rocks with relative higher resistivity values.

The aquifer zones are interpreted as average resistivity values (20–100 Ωm). The homogeneity of the geological formation reflects smooth curves shapes. In the middle area, very low resistivity values (0–5 Ωm) were recorded, indicating the existence of saline water or clayey layers.

2.3.2 Quantitative interpretation

The VES curves are interpreted with $IPI\text{-}_{2WIN}$ software. The program provides an option of inverse modeling for the field curves as well as other options for direct interpretation. In this study, a model for each field curve is proposed based on the type of the curve and its resistivity variations. The program on iteration computes a theoretical curve based on an input model and calibrated with the field curve. The validity of the resulting model is checked against the prevailing lithological and hydrological conditions in the vicinity of the VES point. Some geophysical data could be directly converted to geological term, where other geophysical data require substantial processing before they confirm certain geological indication after interpretation [17].

Available of (64) boreholes, log data descriptions that cover the study area have been used to assist the interpretation of the VES data (Appendix C). Variations in apparent resistivity values were attributed to the inhomogeneity of formations strata.

2.3.2.1 Upstream area

The upstream area started from Showak town including Upper Atbara and Setit dam areas and ended at Wad Elhelew town to the southeast.

In Showak area upstream, the dominant curves are HKH and KH types indicating 7 layers with 3 saturated zones with resistivity values ranging between 2 and 600 Ωm. Near the river Atbara, the Static Water Levels (SWL) were monitored at shallow depth ranging between 3 and 7 m. The recognized conductive zones characterize the upper and lower aquifers and consist of Cretaceous sedimentary formation (**Table 1**).

Around Abbuda camp in upstream area, the resistivity values are ranging between 10 and 600 Ωm. The dominant curves are HKH and KH types indicating 5 layers. The upper aquifer is composed of River Atbara sediments, where the lower aquifer consists of Cretaceous Sandstone.

Layer No.	Depth (m)	Resistivity (Ω.m)	Geological units
1	0–10	2–10	Superficial deposits (composed of silts and fine sands)
2	10–70	60–120	Saturated River Atbara sediments
3	70–90	10–20	Dark clayey layers
4	90–120	30–60	Upper aquifer composed of Cretaceous Sandstone; medium to coarse grained, various in color.
5	120–160	5–10	Clay sequence (gray in color).
6	160–200	30–80	Lower aquifer composed of Cretaceous Sandstone; medium to coarse grained, various in color.
7	> 200	> 600	Basement complex (granite and quartzite)

Table 1.
The interpretation of resistivity curves in Showak area.

In **Table 2**, the static water levels (SWL) observed in boreholes are ranging between 58 and 62 m below ground surface.

At Umm Gargoor area upstream, the dominant curves are HA and KH types signifying 7 layers with resistivity values ranging between 6 and 200 Ωm. The Cretaceous Sandstone layer represents a conductive zone (**Table 3**). In some boreholes drilled inside the refugee camp, the recorded SWL is ranging between 70 and 73 m.

In Wad Elhelew area, the dominant curves are HKH and KH types representing 6 different layers, with resistivity values ranging between 12 and 300 Ωm. The water-bearing zones composed of Cretaceous Sandstone and basaltic rocks (**Table 4**). The static water levels (SWL) are ranging between 25 and 41 m below surface.

In Upper Atbara and Setit dam areas, the number of (31) vertical electrical sounding points have been concentrated. The dam area are divided into two parts, which are Gedaref and Kassala areas.

The VESs were distributed as gird points in the Gedaref area with easting interval of 500 m and northing interval of 800 m; the AB/2 separation reached 700–800 m. The resistivity values are ranging between 2 and 600 Ωm. The dominant curves are detected as HA and A types, representing 6–7 layers. The HA curve shows apposite conductive zone, while the A curve type shows the transition of resistivity values to a

Layer No.	Depth(m)	Resistivity (Ω.m)	Lithological descriptions
1	10	10–30	Superficial deposits
2	10–70	40–70	Upper aquifer consist of saturated River Atbara Sediments
3	70–110	100–300	Basaltic rocks (saturated fractured zone at the top, very hard and dry by depth)
4	110–140	20–30	Lower Aquifer (Cretaceous Sandstone), medium to coarse grained.
5	> 140	> 600	Basement complex (granite and quartzite).

Table 2.
The interpretation of resistivity curves in Abbuda area (near to the River Atbara).

Layer No	Depth(m)	Resistivity (Ω.m)	Geological terms
1	0–5	6–10	Black cotton soil
2	5–25	18–20	Dry sand
3	25–50	20–32	Clay, multicolor.
4	50–100	30–50	Basaltic rock, very hard to penetrate.
5	`100–120	35–70	Aquifer zone (Cretaceous Sandstone, medium to coarse grained, various in color)
6	120–200	40–90	Mudstone overlain the basement complex
7	> 200	> 200	Basement complex (granite and quartzite)

Table 3.
The interpretation of resistivity curves in Umm Gargoor area.

Layer No.	Depth (m)	Resistivity (Ω.m)	Lithological description
1	0–30	30–60	Superficial deposits
2	30–38	12–25	Confining clayey layers
3	38–90	60–70	Cretaceous Sandstone (coarse grained, partly saturated).
4	90–120	110–130	Weathered/fractured saturated basaltic rocks.
5	90–120	60–80	Aquifer of Cretaceous Sandstone (medium to coarse grained).
6	> 120	> 300	Intruded basaltic rocks (dry and hard).

Table 4.
The interpretation of resistivity curves in Wad Elhelew area.

higher one, which they designate compacted layer on the basement complex, interpreted as compacted sandstone formation (**Table 5** and **Figure 5**).

The VESs were distributed in gird points in Kassala area, with easting interval of 750 m and northing interval of 500 m. The measured resistivity values are generally ranging between 10 and 600 Ωm. The A, HA, HKH, and KH curve types are common

Layer No	Depth(m)	Resistivity (Ω.m)	Geological term
1	0–10	2–10	Superficial deposits (silts and fine sands)
2	10–25	60–120	Cretaceous Sandstone (dry and compact)
3	25–40	10–20	Clay materials
4	40–120	30–60	Aquifer zone of Cretaceous Sandstone (Medium to coarse grained).
5	120–160	5–10	Mudstone and shale
6	> 160	> 600	Basement complex

Table 5.
The interpretation of resistivity curves in the dam area-Gedaref area.

DOI: http://dx.doi.org/10.5772/intechopen.1007086

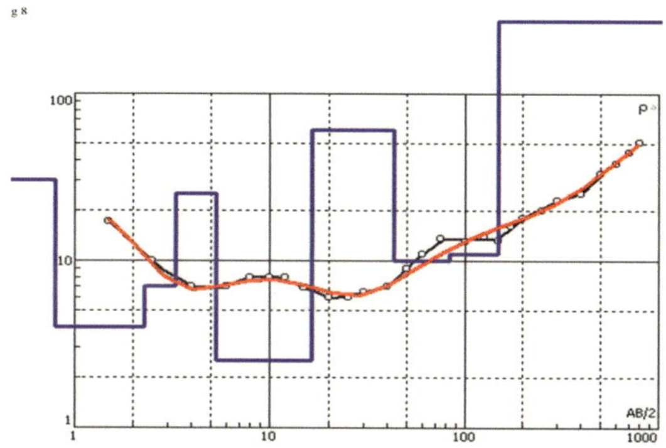

RMS=7.07%

N	1	2	3	4	5	6	7	8	9	
ρ	30	4	7	25	2.5	60	10	11	2686	
h	0.8	1.5	1	2	11	27	40	65		
d	0.8	2.3	3.3	5.3	16.3	43.3	83.3	148		
Alt	-0.8	-2.3	-3.3	-5.3	-16.3	-43.3	-83.3	-148.3		

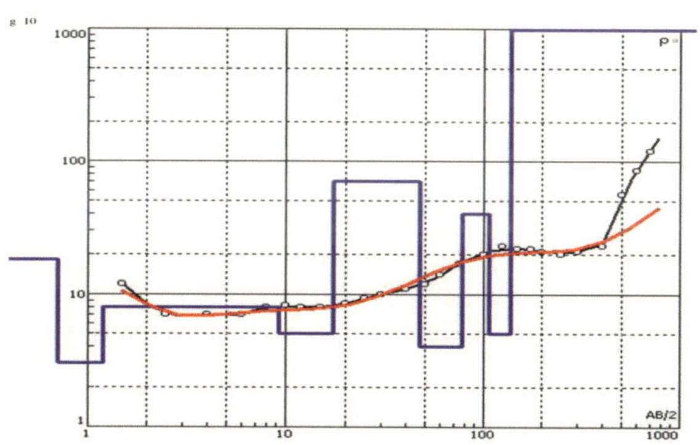

RMS=33.6%

N	1	2	3	4	5	6	7	8	9	
ρ	18	3	8	5	70	4	40	5	4482	
h	0.7	0.5	8	8	30	30	30	30		
d	0.7	1.2	9.2	17.2	47.2	77.2	107	137		
Alt	-0.7	-1.2	-9.2	-17.2	-47.2	-77.2	-107.2	-137.2		

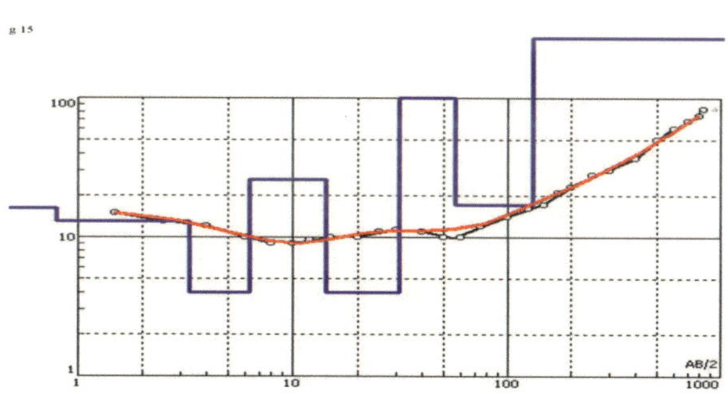

RMS=4.87%

N	1	2	3	4	5	6	7	8		
ρ	16	13	4	26	4	100	17	4314		
h	0.8	2.5	3	8	17	25	75			
d	0.8	3.3	6.3	14.3	31.3	56.3	131			
Alt	-0.8	-3.3	-6.3	-14.3	-31.3	-56.3	-131.3			

Figure 5.
Some typical measured resistivity curves in Gedaref area (Upper Atbara dam area).

(**Tables 4** and **5**). The KH/HKH curve types indicating 5–7 layers with conductive zones of apparent resistivity ranges between 20 and 90 Ωm (**Table 6** and **Figure 6**).

In the dam area, the VES data reflect the resistivity variations in the subsurface geological strata. The upstream area represents a sub-basin in which thick Cretaceous sedimentary rocks were preserved (**Table 7** and **Figures 5** and **6**). The SWL as observed in the nearby observation wells is ranging between 45 and 57 m below the ground surface.

Based on (**Table 7**) the apparent resistivity at depths of 100 and 120 m for Gedaref area, it can be interpreted as that:

- Major Fault structure trends from southwest to north east through middle area.
- The aquifer is composed of sandstone at depth range between 80 and 110 m, and there is no indication for the occurrence of basaltic rocks through the sequences.
- The conductive zone is represented by resistivity values range between 30 and 60 Ωm.
- The high values of apparent resistivity may indicate the existence of hard formation or basement complex at depth of 160 m. The conductive zones occurred at seven points as displayed in **Table 8**.

Based on (**Table 7**) the apparent resistivity at depth of (100,120) m for Kassala area, it can be interpreted as that:

- The high values of apparent resistivity in eastern side at VES (14) may indicate intrusion of thick basaltic formation.
- The conductive zone is represented by values range between 10 and 30 ohm\m.
- The conductive zones occur at nine points in the area as displayed in **Table 9**.
- The high values of apparent resistivity may indicate hard formation or basement complex.

Layer No.	Depth(m)	Resistivity (Ω.m)	Lithological description
1	0–1	10–30	Superficial deposits
2	1–5	2–5	Clay, sticky and compact
3	5–20	100–450	Dry friable formation
4	20–30	15–30	Clay
5	30–110	100–300	Intrusive basaltic rocks (Fractured and saturated at the top and very hard and dry below)
6	110–140	20–30	Aquifer zone (Cretaceous Sandstone, medium to coarse grained, various in color)
7	> 140	> 600	Basement complex (granite and quartzite)

Table 6.
The interpretation of resistivity curves in the dam area–Kassala area.

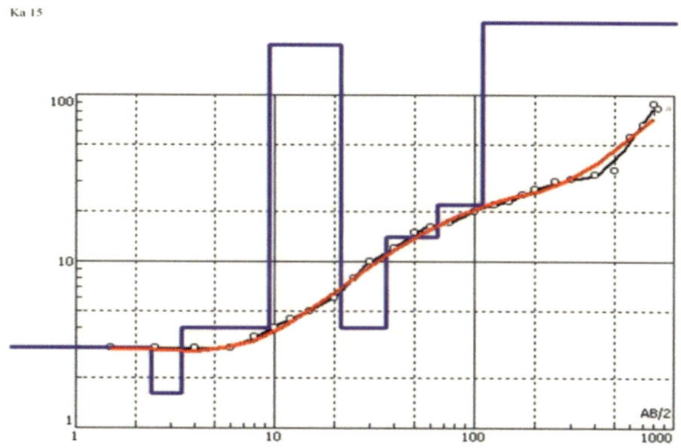

RMS=6.55%

N	1	2	3	4	5	6	7	8		
ρ	3	1.6	4	200	4	14	22	5424		
h	2.4	1	6	12	15	29	44			
d	2.4	3.4	9.4	21.4	36.4	65.4	109			
Alt	-2.4	-3.4	-9.4	-21.4	-36.4	-65.4	-109.4			

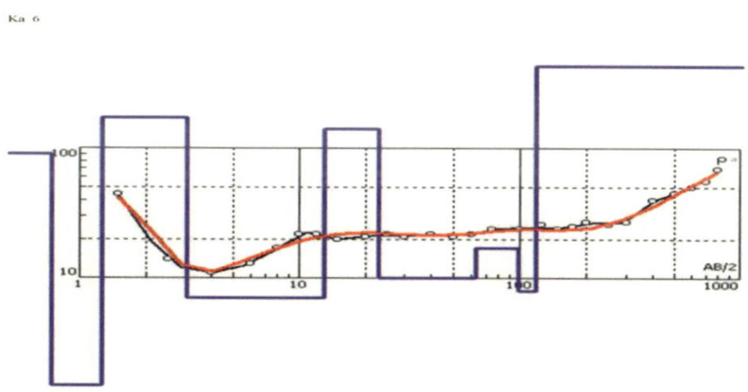

RMS=7.06%

N	1	2	3	4	5	6	7	8	9	
ρ	90	1.5	170	7	140	10	17	8	432	
h	0.75	0.5	1.8	10	10	40	35	20		
d	0.75	1.25	3.05	13.1	23	63	98.1	118		
Alt	-0.75	-1.25	-3.05	-13.05	-23.05	-63.05	-98.05	-118.1		

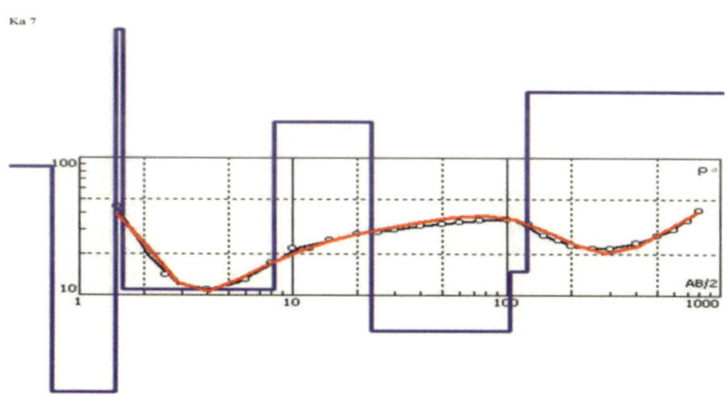

RMS=6.16%

N	1	2	3	4	5	6	7	8		
ρ	85.2	1.95	3825	10.9	181	5.5	15	300		
h	0.75	0.717	0.106	6.57	15	80	20			
d	0.75	1.47	1.57	8.14	23.1	103	123			
Alt	-0.75	-1.467	-1.573	-8.143	-23.14	-103.1	-123.1			

Figure 6.
Some typical measured resistivity curves in Kassala area (Setit dam area).

Gedaref study area			Kassala study area		
VES points	Resistivity (Ωm) at 100 m depth	Resistivity (Ωm) at 120 m depth	VES points	Resistivity (Ωm) at 100 m depth	Resistivity (Ωm)at 120 m depth
1	30	33	1	35	42
2	27	30	2	40	49
3	35	40	3	46	52
4	30	35	4	44	47
5	44	47	5	60	50
6	62	96	6	34	40
7	32	52	7	20	23
8	24	44	8	48	54
9	30	40	9	46	54
10	20	22	10	50	60
11	35	47	11	40	52
12	32	38	12	46	47
13	32	40	13	44	38
14	50	56	14	40	42
15	30	38	15	30	44

Table 7.
The apparent resistivity at 100, 120 m depths for Gedaref and Kassala areas.

Site No	Coordinates		Recommended depth (meters)	Remarks
10	1,572,728	0811376	137	
11	1,573,436	0811749	136	
12	1,574,134	0812141	156	
2	1,575,178	0811850	141	
7	1,572,500	0811821	143	
8	1,571,789	0811456	148	
9	1,572,017	0811011	154	

Table 8.
Recommended sites for drilling boreholes in Gedaref area (River Atbara).

In comparison the results of the lithological variations and thickness of the aquifers in Kassala and Gedaref areas, (16) out of (31) VES points were recommended for drilling boreholes (**Tables 8** and **9**). The relatively higher resistivity values are attributed to the existence of a thick layer of basalts in Kassala and a thick layer of mudstone in Gedaref areas.

Site No	Coordinates		Recommended depth (meters)	Remarks
4	1,580,610	0818767	129	
6	1,581,960	0820416	118	
10	1,580,982	0817657	121	
11	1,580,700	0816497	123	
12	1,580,261	0816633	127	
13	1,580,467	0817188	150	
14	1,580,927	0818771	151	
2	1,581,105	0820188	147	Weak conductive zone
5	1,580,374	0818066	135	Weak conductive zone

Table 9.
Recommended sites for drilling boreholes in Kassala area (River Setit).

2.3.2.2 Middle area

The middle part is the large part of the study area; it extended from Al Shagarab villages to the north of Showak town. The recorded resistivity values are ranging between 3 and 400 Ωm. The common curves are KH, KHA, and A types. The KH and KHA curves indicate 4–5 layers as saturated water horizons. Relatively thick River Atbara sediments with different grain size occur in considerable thickness below the surface representing a sedimentary basin. The high resistivity value was represented in A curve type indicating the existence of basaltic rocks overlaying the basement rocks. Away from the River Atbara course, dry and compacted stratum of basaltic rocks occurred instead of River Atbara sediments (**Tables 10** and **11** and **Figure 7**). The SWL in the middle area, ranging between 9 and 10 m close to the River Atbara and 20–25 m away from the river course.

2.3.2.3 Downstream area

The downstream part is located between the Shagarab villages and Khashm El Girba dam at the north. The resistivity values are ranging between 10 and 500 Ωm. The A curve type is dominant, indicating 3–4 layers. The low resistivity values are referring to clayey layers within the superficial deposits. The underlain basaltic rocks

Layer No.	Depth (m)	Resistivity (Ωm)	Geological term
1	0–10	10–300	Superficial deposits (fine sands & silts).
2	10–20	1–20	Clayey layer.
3	20–60	20–70	River Atbara sediments (medium to coarse sand & gravels).
4	60–120	100–300	Basaltic rocks (Fractured/saturated in top, hard & dry below)
5	> 120	> 500	Basement complex

Table 10.
The dominant curve types in the middle area (Near the River Atbara).

Layer No.	Depth (m)	Resistivity (Ωm)	Geological term
1	0–10	3–60	Clay
2	10–30	20–100	Dry sand
3	30–140	120–200	Intrusive basaltic rocks, Dry and very hard to penetrate
4	> 140	> 300	Basement complex (granite and quartzite)

Table 11.
The dominant curve types in the middle area (away from River Atbara).

Um ood 1

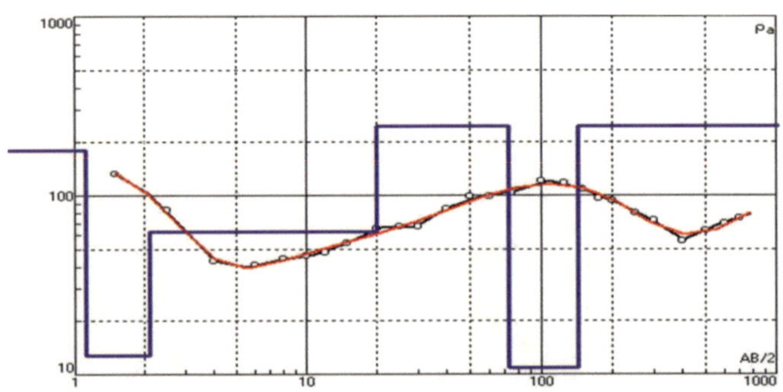

Error = 3.57%

N	1	2	3	4	5	6
ρ	180	12.7	63.1	242	10.9	244
h	1.12	1	18	52.4	70.7	
d	1.12	2.12	20.1	72.5	143	
Alt	-1.116	-2.119	-20.09	-72.52	-143.3	

alruhal2

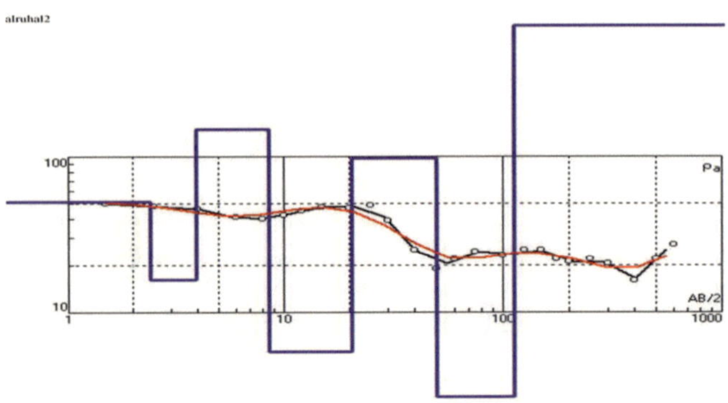

Error = 7.38%

N	1	2	3	4	5	6	7
ρ	51.6	16.3	149	5.59	97.3	2.84	1017
h	2.4	1.53	4.55	12.2	29.1	62.9	
d	2.4	3.93	8.48	20.6	49.7	113	
Alt	-2.398	-3.927	-8.48	-20.64	-49.75	-112.6	

Figure 7.
Some typical measured resistivity curves in middle area.

show relatively higher resistivity values over the basement rocks. The downstream area considered as none water-bearing formation; in some seats in the downstream area near River Atbara, a thin layer of River Atbara sediments is recorded as limited water zone. The fractures/joints basaltic rocks represent a saturated zone within the river basin near the river bank indicating direct recharge from the river (**Tables 12** and **13** and **Figure 8**). The SWL in the downstream area ranges between 6 and 8 m depths as it observed from the existing wells data.

2.4 Geo-electrical sections in the study area

Eight geo-electrical sections were constructed using the results of the interpretation of the VES results, which correlated with the existed boreholes data. The geo-electrical sections were formed in N-S and E-W directions. Two geo-electrical sections are located on the upstream area, three geo-electrical sections situated on the middle area, and two sections located on the downstream area. One geo-electrical sections extended along the study area in N-S direction from the downstream to upstream areas (**Figure 9**).

2.4.1 Geo-electrical section No.1

The section involves VES105, VES99, VES96, VES95, VES163, VES143, VES148,VES 149, VES61,VES60,VES25,VES56,VES39, and VES 37 (**Figure 9**). This section started from Jebel Karkora and terminated at Alfashaga area trending in N–S direction. It contains 7 layers; from the top to bottom, superficial deposits with thickness ranging between (2–20 m) with resistivity values range from 2 to 30 Ωm. The second layer consists of dry sand with thickness ranging between 5 and 20 m with resistivity values range from 18 to 20 Ωm. The third layer is clay with thickness ranging between 1 and 10 m with resistivity values range from 2 to 6 Ωm. The fourth layer is basaltic rocks, which started from Jebel Karkora in Kassala area and terminated at VES99 in Gedaref area (**Figure 10**). The layer thickness is ranging between 50 and 80 m with resistivity values range from 100 to 300 Ωm. The fifth layer represents the aquifer zone that consists of sandstone with thickness ranging between 30 and 40 m, and resistivity values range from 30 to 80 Ωm. This sandstone layer sometimes occurred dry and compact at the surface near the Upper Atbara River

Layer No.	Depth (m)	Resistivity (Ω.m)	Geological unit
1	0–10	10–300	Superficial deposits (consist of fine sand and silt).
2	10–30	30–40	River Atbara sediments (aquifer formation).
3	30–100	60–80	Fractured basaltic
4	>100	> 100	Basement complex

Table 12.
The interpreted geo-electrical section in downstream area (Near the River Atbara).

Layer No.	Depth (m)	Resistivity (Ω.m)	Geological term
1	0–10	10–75	Clay, black in color.
2	10–60	100–300	Intrusive basaltic rocks (dry and very hard to penetrate)
3	> 60	> 500	Basement complex (granite and quartzite)

Table 13.
The interpreted geo-electrical section in downstream area (away from the River Atbara).

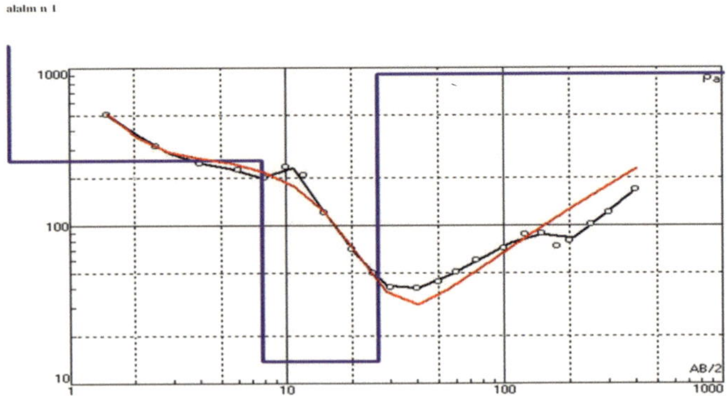

Error = 19.5%

N	1	2	3	4							
ρ	1388	257	13.8	917							
h	0.503	7.26	18.7								
d	0.503	7.76	26.4								
Alt	-0.5025	-7.762	-26.42								

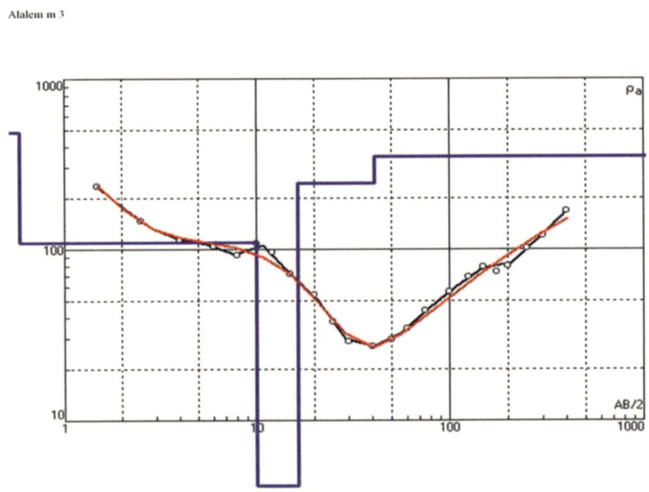

Error = 7.04%

N	1	2	3	4	5						
ρ	487	110	4.18	245	348						
h	0.59	9.35	6.38	24.5							
d	0.59	9.94	16.3	40.8							
Alt	-0.5902	-9.944	-16.33	-40.82							

Figure 8.
Some typical measured resistivity curves in the study area (Downstream area).

bank with thickness ranging between 2 and 30 m, and resistivity values range from 60 to 120 Ωm. The sixth layer is detached layer of mudstone occurred at the middle part of the cross section with thickness range from 40 to 80 m; the resistivity values range from 40 to 90 Ωm. The last unit is the basement rocks located at 120–170 m depth; the resistivity values range between 200 and 600 Ωm (**Figure 10**).

Generally, the geological units in this section are structurally controlled. Three points (VES105, VES99, and VES96) are recommended for drilling boreholes; one point of them (VES 149) was confirmed by drilling.

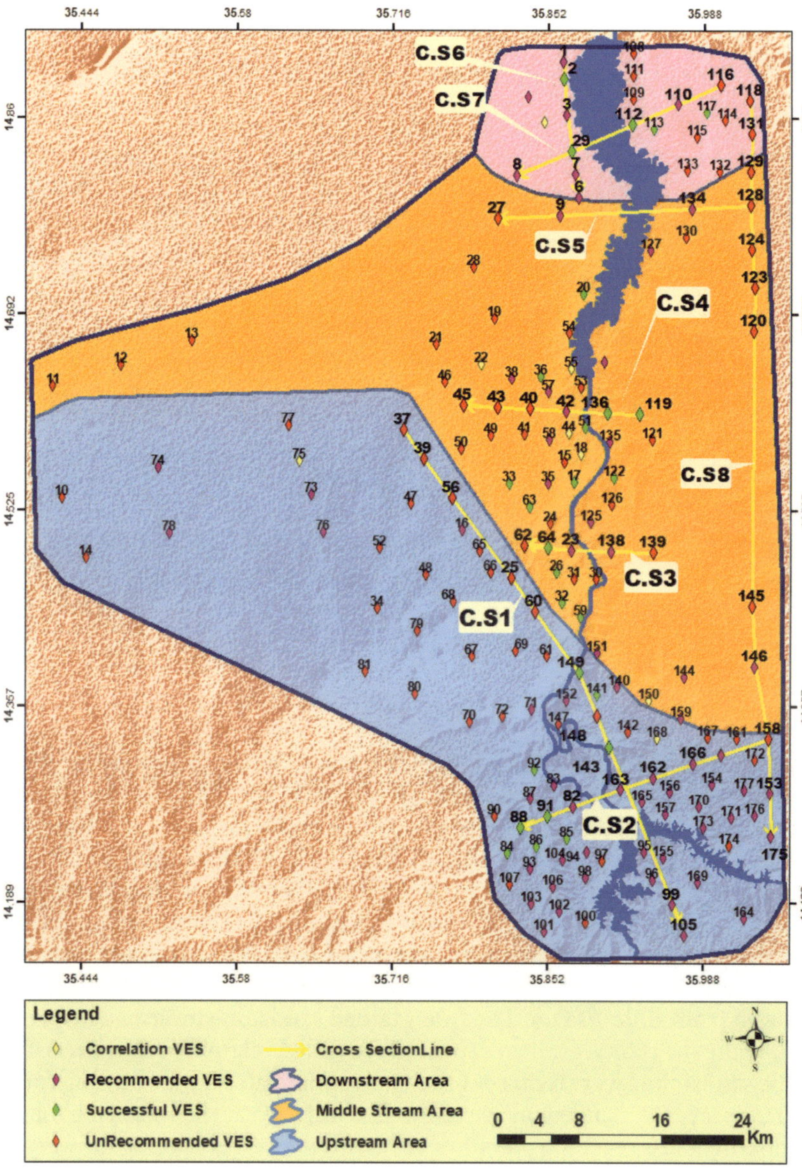

Figure 9.
Location map of the geo-electrical sections in the study area.

2.4.2 Geo-electrical section No.2

The geo-electrical section was a buildup from VES158, VES160, VES166, VES162, VES163,8 VES2, VES 91, and VES88. The constructed section started from the eastern bank of the River Setit (Kassala area) and terminated at the western bank of River Atbara (Gedaref area) in E–W direction (**Figure 9**). It consists of 7 layers; the top layer is a form of superficial deposit with 5 and 20 m) thickness, with resistivity values ranging between 2 and 30 Ωm. The top layer was underlain by a thin layer of dry sand portray thickness ranging between 6 and 14 m with resistivity values ranging from 100 to 450 Ωm. The sandy layer was observed on the eastern bank of the River

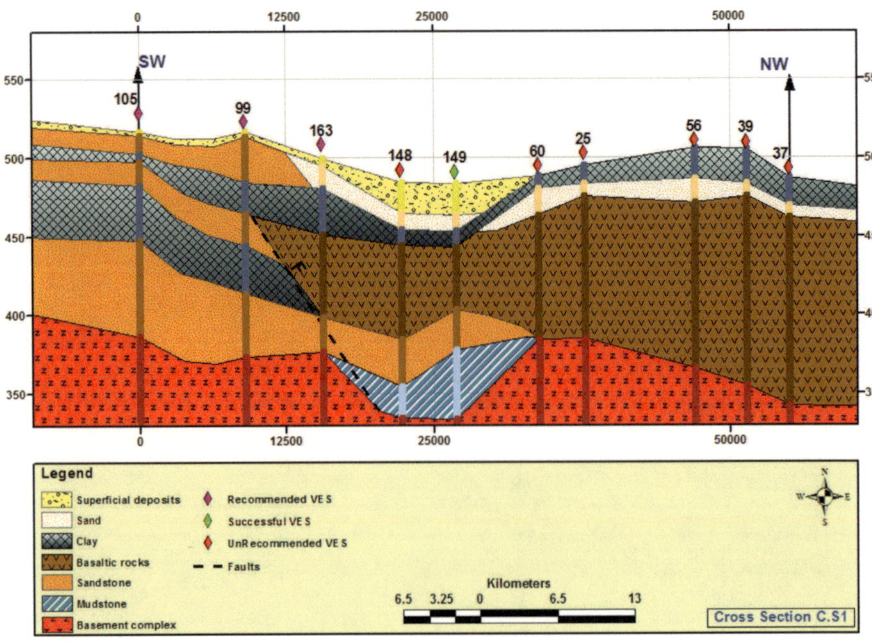

Figure 10.
Cross section No.1 located at the upstream part of the study area.

Atbara (Kassala area). A clayey confining layer underlies the dry sandy layer of 10 and 50 m thickness, with resistivity values ranging from 2 to 30 Ωm. The basaltic rocks is expected to underlie the clayey layer with thickness ranging between 20 and 90 m, with resistivity values ranging from 100 to 300 Ωm. The thickness of basaltic layer increased on the eastern bank of the River Atbara (Kassala area) and moved out at the Gedaref area, at the western bank of the River Atbara. The saturated zone is of relatively low resistivity compared to that unsaturated zone. The aquifer zone is expected below the basaltic layer, form of sandstone with 20–65 m thickness, and resistivity values ranged from 10 to 70 Ωm. The fine-grained sandstone indicated low resistivity values, while the coarse grained sandstone revealed relatively higher resistivity values. The sandstone layers represent the main appreciable water bearing formation. The mudstone layer is mainly concise to the Gedaref area at the western bank of River Atbara with thickness reaching 30–40 m of resistivity values ranging between 1 and 5 Ωm. The basement complex represents the deepest unit situated between 120 and 160 m depths of resistivity values ranging from 600 to 2100 Ωm.

The conducted (WES) points in this section were considered as promising sites for drilling boreholes except the point at VES 158, which is not recommended for drilling in which the thick basaltic layer is expected (**Figure 11**). The drilling confirmed the groundwater potentiality at VES 91 and VES 88 locations. This section confirmed that the western bank of the River Setit was uplifted in which the riverbed is highest by 4 meters compared to River Atbara bed.

2.4.3 Geo-electrical section No.3

This cross section was based on VES139, VES138, VES23, VES64, and VES62 that extended in E–W direction (**Figure 9**). The section shows six layers, starting from a thin layer of superficial deposits at the top with thickness varies between 5 and 15 m,

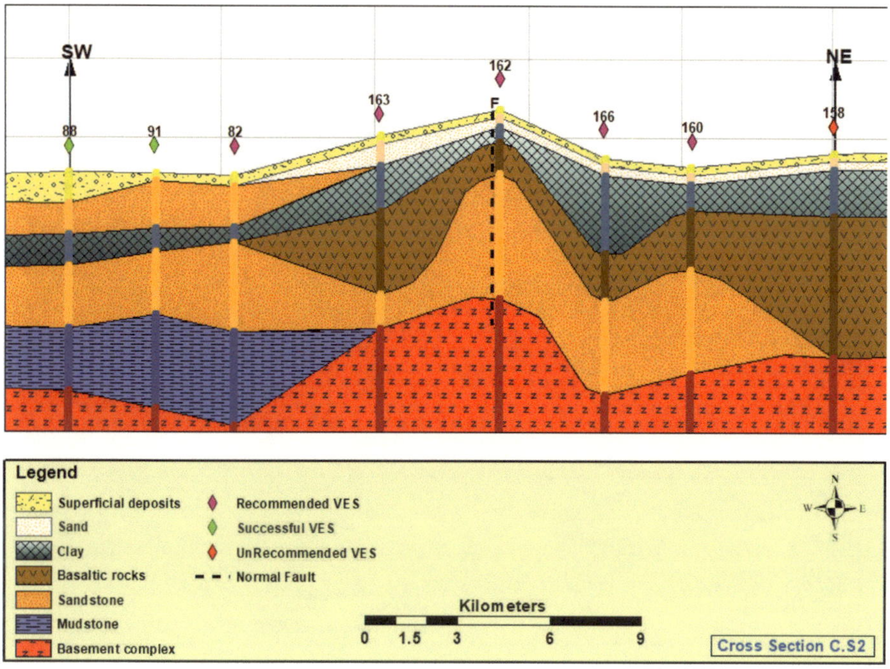

Figure 11.
Cross section No.2 located at the upstream part of the study area.

with resistivity values ranging from 10 to 300 Ωm. A confining layer of clay with thickness ranging between 5 and 25 m with resistivity values ranging from 1 to 20 Ωm underlies the top superficial deposits. A dry sandy layer occurred at the right and left sides of the cross section and missed in the middle part of the cross section display layer thickness ranging between 10 and 30 m, with resistivity values ranging from 20 to 100 Ωm. A layer interpreted as River Atbara sediments, partly saturated (gravels, medium to coarse sand), extended within the River Atbara bank and represents the main aquifer in the middle stream area. Its thickness reaches 40 m, and resistivity values range from 20 to 70 Ωm. A very thick layer of basaltic rocks occurred of 60 and 90 m thickness of resistivity values ranging from100 to 300 Ωm. The depth of the basement complex is 140 m in the west direction (**Figure 12**). The recommended points (VES138, VES23, and VES 64) were located near River Atbara within River Atbara sediments aquifer, which they confirmed by successful drilling. VES 62 is not recommended for drilling that is located far from the River Atbara and expected to be situated within the basaltic rocks extension.

2.4.4 Geo-electrical section No.4

The geo-electrical section contains VES119, VES136, VES42, VES40, VES43, and VES45 crossing the main River Atbara in the middle stream area in E-W direction (**Figure 9**). This section reveals six formational units; the top layer confirms superficial deposits with 10 to 30 m thickness. The resistivity values of this layer, ranging between 10 and 300 Ωm, referred to the existence of dry sand; the layer terminated at VES40. Clayey layer underlies the top layer with 13–40 m thickness, and resistivity values range between 3 and 60 Ωm. On the western and eastern sides of the cross

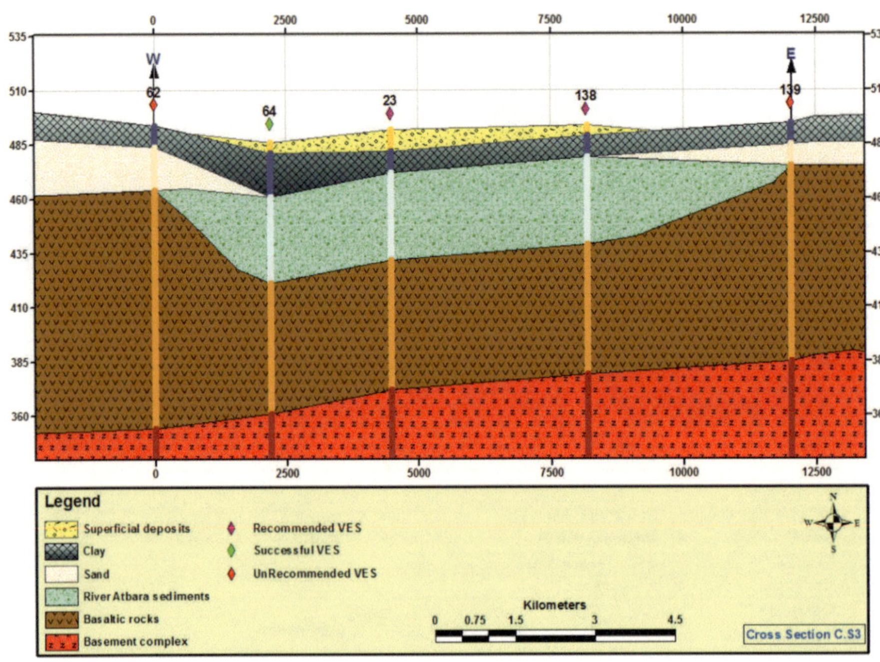

Figure 12.
Cross section No. 3 located at the middle part of the study area.

section, a thin dry sandy layer occurred and disappeared somewhere at the middle of the section. Partially saturated layer underlies the sandy layer composed of River Atbara sediments and reaches 40 m thickness, with resistivity values ranging from 20 to 70 Ωm. A thick basaltic layer underlies the River Atbara sediments and rests on the basement rocks. It reaches 60–110 m thickness, and resistivity values range from 100 to 300 Ωm. The depth to basement rocks is expected to be between 113 and 160 m of resistivity values ranging from 300 to 500 Ωm (**Figure 13**).

The River Atbara sediments are characterized the aquifer zone located at stations (VES136, VES42); the saturated fracture zone of the basaltic rocks is detected at station VES119 (**Figure 13**). VES119 and VES136 are confirmed by successful drilling, resulting good yield. VES40, VES43, and VES45 are not recommended for drilling located out of the aquifer zone, while VES42 represents the promised site for drilling.

2.4.5 Geo-electrical section No.5

This section includes VES128, VES134, VES9, and VES27, oriented in E-W direction (**Figure 9**). It indicates four layers; the upper layer in the section was dominated by clayey layer, of resistivity values ranging between 1 and 20 Ωm and thickness ranging between 10 and 20 m. A dry sandy layer underlies the upper layer with resistivity values ranging between 18 and90 Ωm; the thickness of this layer ranges between 10 and 30 m. The layer consists of injected basaltic rocks underlying the sandy layer. The resistivity values of the intrusive layer range between 100 and 300 Ωm, with thickness ranging between 40 and 65 m. These basaltic rocks are fractured and saturated at the top level; it poses very hard and dry in the lower part. The basement complex occurred as the lowest unit in this section with resistivity values ranging between 300 and 500 Ωm. The depth to the basement complex reaches above 80 m. The basement

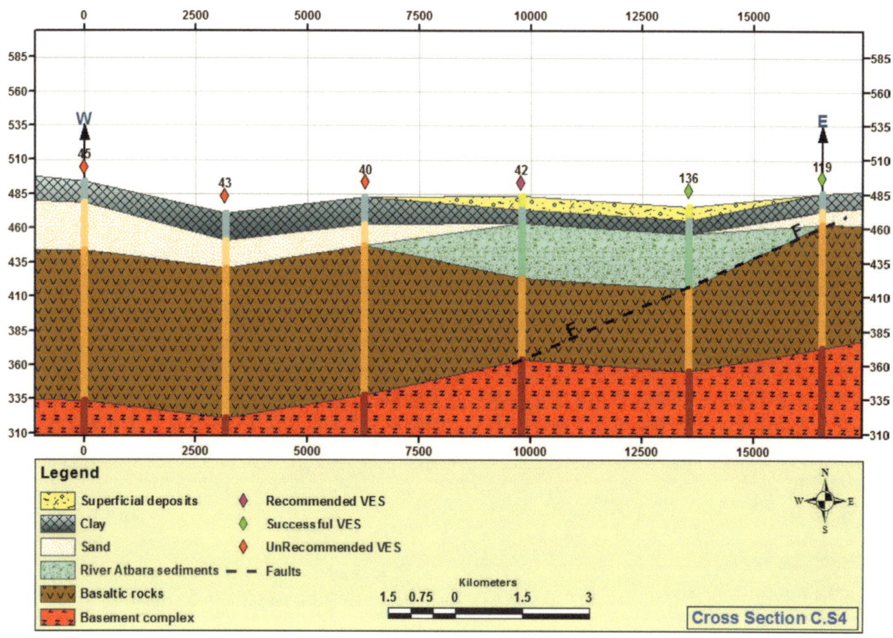

Figure 13.
Cross section No.4 located at the middle part of the study area.

complex occurred at a shallower level at the eastern bank of the river Atbara compared with the western bank (**Figure 14**).

In this section, VES134 and VES9 are recommended for drilling boreholes, which are located on the saturated fractured basaltic zones attached to the river. VES128 and VES27 are not recommended for drilling because they are situated away from the saturated zone.

2.4.6 Geo-electrical section No.6

The section is constructed on VES1, VES2, VES3, VES29, VES7, and VES6 in N-S direction (**Figure 9**). It is composed of four layers, superficial deposits with resistivity values ranging between 12 and 75 Ωm and thickness ranging from 5 to 10 m. The River Atbara sediments which are represent a saturated water bearing formation with resistivity values range from 30 to 40 Ω.m, and 10–20 m thick. The saturated fractured and jointed basaltic rocks vary in resistivity values ranging from 60 to 80 Ωm with thickness varying from 10 to 75 m. The basement complex is of relatively high resistivity values ranging from 100 to 500Ωm and depths varying from 30 to 100 m. The depth to the basements increases from south to north direction, where it crops out on the eastern bank of the River Atbara bed near Khashm El Girba dam at the north (**Figure 15**).

All VES points of this section are located on the aquifer zone consisting of River Atbara sediments. The drilling confirmed VES2 and VES29, which evidence saturated water-bearing zone.

2.4.7 Geo-electrical section7

VES116, VES110, VES112, VES29, and VES8 were used to build the geo-electrical Section 7. The section is situated on the downstream area extended in E–W direction

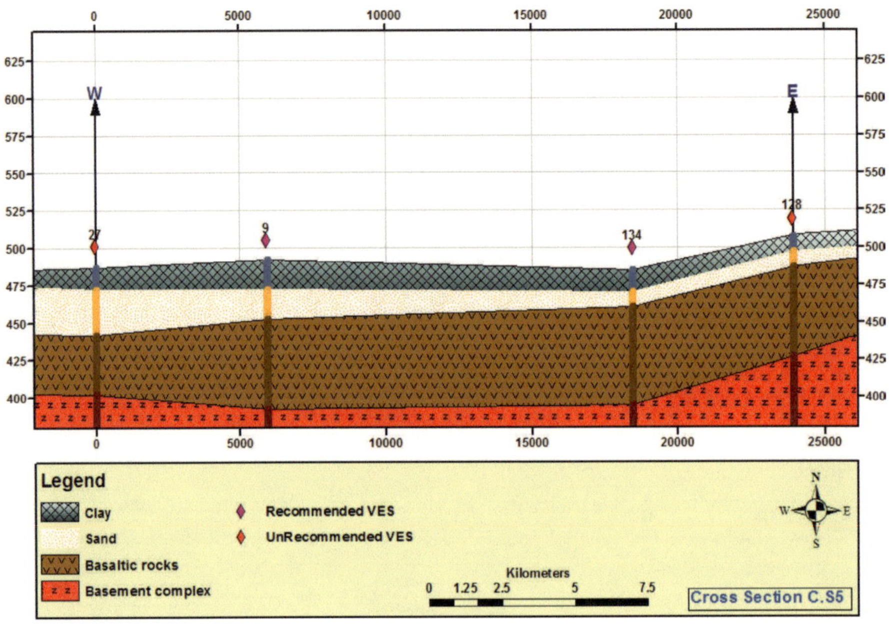

Figure 14.
Cross section No.5 located at the middle part of the study area.

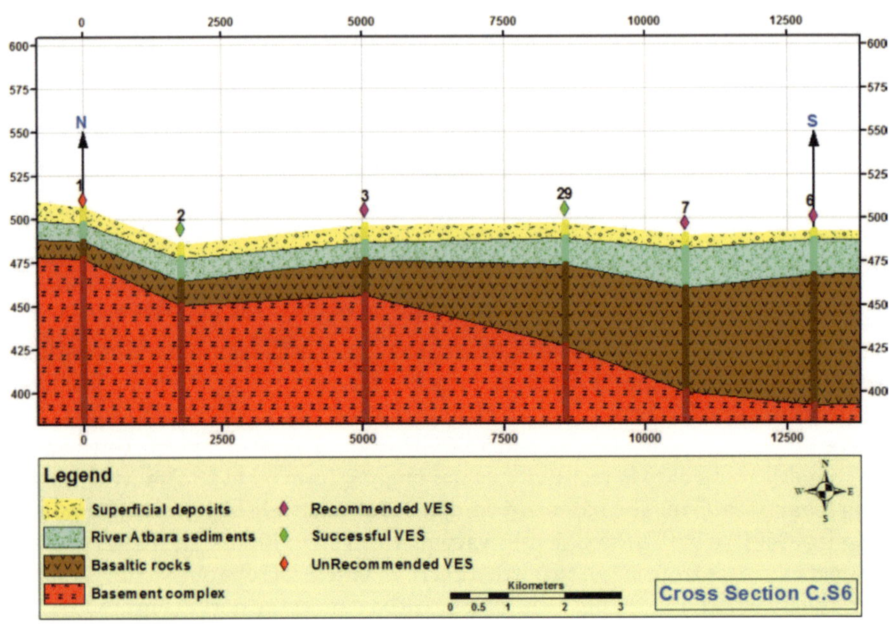

Figure 15.
Cross section No.6 is located at the downstream part of the study area.

(**Figure 9**). The section revealed five geological sequences; superficial deposits detected by (VES112 & VES29) and displayed near the river. The resistivity values of this layer range from 10 to 300 Ωm with 10 m thickness. The low resistivity values within the formation represent clay layers, while the relatively higher resistivity values indicate dry sandy layers. VES116, VES110, and VES8 indicate clayey layer located away from the river with resistivity values ranging between 10 and 75 Ωm and 10 m thickness. The top layers overlay the aquifer formation composed of River Atbara sediments that occurred near to the river. VES112 and VES29 define the saturated formation at the middle part of the cross section. The recorded resistivity values of the River Atbara sediments range from 30 to 40 Ωm with 10–15 m thickness. A sequence of basaltic rocks is expected to rest on basement complex rocks with resistivity values ranging from 60 to 300 Ωm and 20–50 m thickness. The top part of these basaltic rocks is fractured and saturated at shallow levels (near the recharge source) and very hard and dry below. The basement complex rocks represent the bottom unit of the section. The resistivity values range between 100 and 500 Ωm; the depth to the basement reaches 40 m near the river and 70 m away from the river (**Figure 16**).

The geo-electrical section evidences that all VES points are recommended for shallow drilling, except at the VES 116 location, which is situated out of the aquifer zone.

2.4.8 Geo-electrical section No.8

This section constructed along the downstream area, middle area, and upstream area, in the N–S direction, included VES118, VES131, VES129, VES128, VES124, VES123, VES120, VES145, VES146, VES158, VES153, and VES175 (**Figure 9**). The sequences in this section represent five layers: clayey layer, sandy

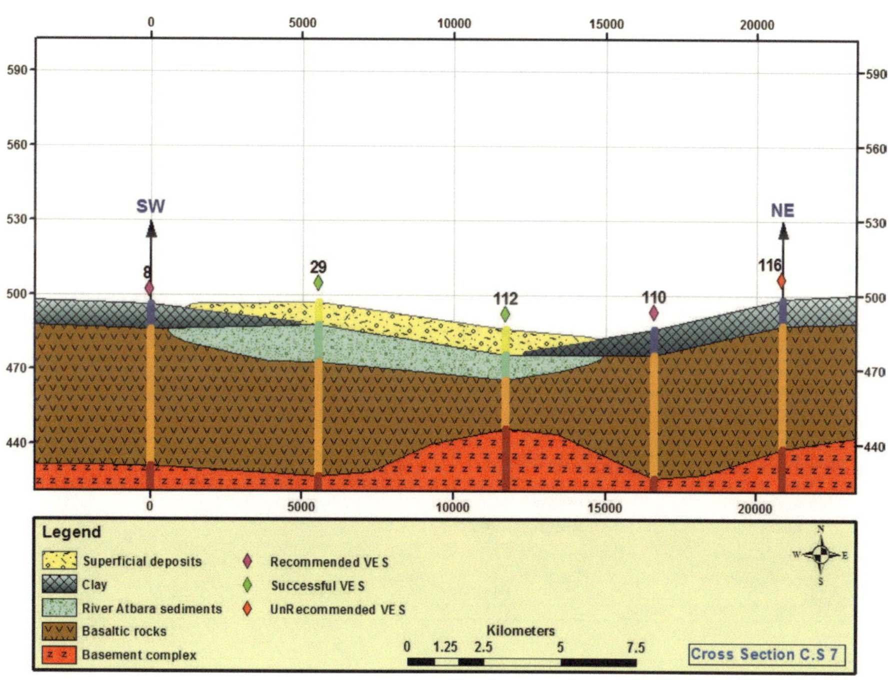

Figure 16.
Cross section No.7 located at the downstream part of the study.

layer, intrusive basaltic layer, sandstone layer, and the basement complex. The resistivity values of clay range from 6 to 75 Ωm with 5–20 m thickness. The resistivity values of the sandy layer range from 18 to 100 Ωm with 6–35 m thickness. This sandy layer was missed in VES131 and VES118 as a result of structural affect. The resistivity values of the basaltic rocks range from 60 to 300 Ωm with 80–115 m thickness. The resistivity values of sandstone range from 20 to 80 Ωm with 11–40 m thickness. This layer was preserved by faults and occurred as discontinued strata that terminated at VES175 and VES153. The southern part of this section is expected as a highly potential zone. The sandstone layer is expressing as a very thin layer at the VES158 location (not recommended for drilling). The resistivity values of the basement rocks range between 100 and 600 Ωm with 110 and 160 m depth (**Figure 17**).

Generally, only the southern part of this section has potential for groundwater occurrence, while the remaining parts of the profile are located in unsaturated zone and not recommended for drilling boreholes.

2.5 The hydro-geophysical investigation findings

The water potentiality map (**Figure 18**) was constructed based on the results of the (177) vertical electrical sounding. The main findings can be summarized as:

- (09) Points were conducted at productive boreholes used for calibration purposes.
- (88) Points are evident as promising drilling sites.

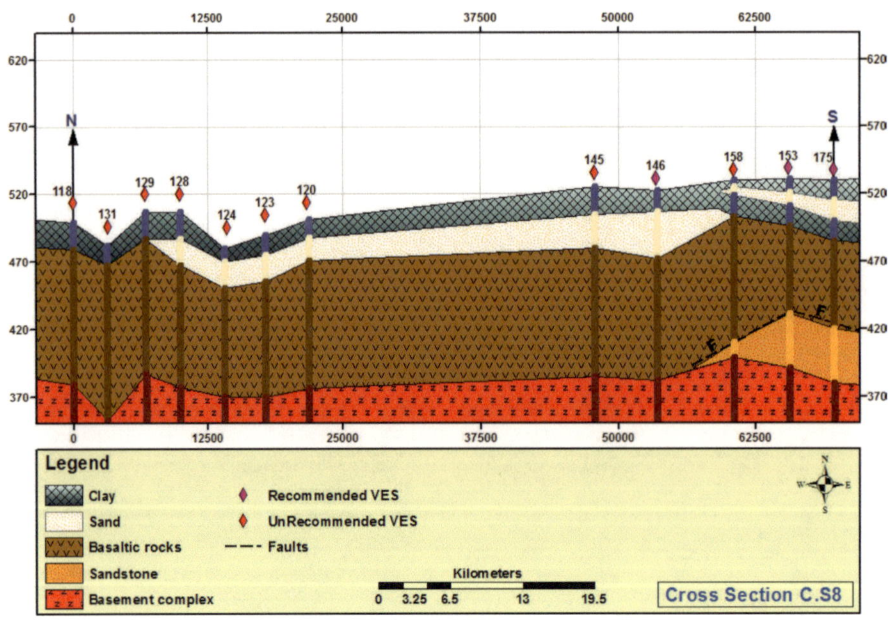

Figure 17.
Cross section No.8 extended from upstream to downstream area.

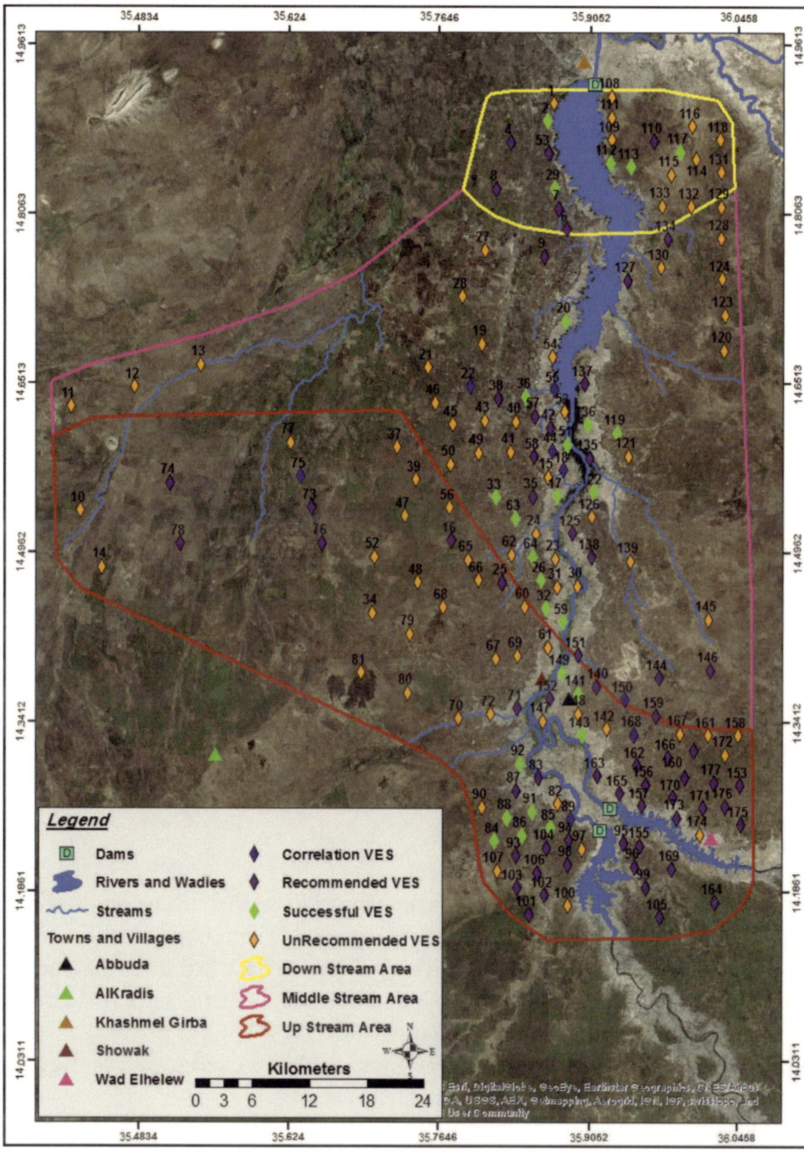

Figure 18.
Water potentiality map (Confirm by drilling data).

- (28) Points were confirmed by drilling.
- (80) Points are not recommended for drilling after data interpretation.

The aquifer in the study area seems to be wider at the upstream area and gets narrower towards the downstream area. The common promising and productive zones consist of:

- The existence of the River Atbara sediments, extended along the River Atbara from the upstream area to downstream area, except on the eastern part of the downstream area near Khashm el Girba dam.

- The existence of the Cretaceous sandstone, which occurs in the upstream part area including Wad Elhelew area, Upper Atbara/Setit dam's area, Showak area, and Um Gargoor area.

- The existence of the saturated fractured basaltic rocks, which covered the area attached to River Atbara banks.

Generally, the VES data indicate that:

- The aquifer thickness of River Atbara sediments is controlled by the lateral and vertical sedimentary facies changes and with the undulations of the basement complex configuration.

- The thickness of the Cretaceous sandstone aquifer zone is affected by the thickness of the intrusive basaltic rocks, thickness of mudstone, and the undulations of the basement complex.

- The saturation of the basaltic rocks is effected by the distance from the river as a recharge source, which is more saturated near River Atbara.

3. Conclusions

The current investigations concerned the Upper River Atbara watershed and its groundwater potentialities. The study area is extended from Upper Atbara and Setit dams situated south to Showak town at about 30 km to Khashm El Girba dam near El Girba city to the north. The study area is located in the arid and semiarid zones, with annual rainfall ranging from 425 to 1166 mm.

Geophysical and hydro-geological investigations were used to assess the groundwater resources in the upper River Atbara and Setit.

The geology of the study area composed of the basement complex, sedimentary rocks Cenozoic basalts, River Atbara sediments, unconsolidated Karab formation, and superficial deposits. Major and minor faults represent the major structural phenomena in the study area controlling formations thickness and the courses of the rivers and khors.

Hydro-geologically, the study area is divided into three sectors: upstream, middle, and downstream areas. The upstream area is characterized by three main aquifer zones as groundwater basins at Showak, Wad Elhelew, and Umm Gargoor area locations. The occurrences of the Cretaceous sandstone layers characterized the aquifer in the upstream area. The aquifer system in the middle area started from River Atbara and extended under the Karab formation. It is bounded by the plain area; the width of the aquifer is extended between 3 and 5 km on each side of the river banks. The aquifer zones in the downstream area do not exceed 1.5 km on the western bank and 1 km on the eastern bank of River Atbara.

Some similarity between the middle and downstream areas: in both areas, the aquifer is composed of River Atbara sediments and fractured basaltic rocks portray different thickness, aquifer zones, and well yield.

The average depth to the basement complex (BC) in the upstream area ranges between 120 and 200 m. The average depth to the basement rocks in the middle area ranges between 90 and 150 m, while it is shallower in the downstream area and ranges between 40 and70 m below ground surface.

Acknowledgements

The authors sincerely thank the Groundwater & Wadies, Kassala and Gedaref offices for their great helps during this research. Moreover, they wish to thank Takweel For Water Investments Company for support during the field work.

Conflict of interest

The authors declare that there is no any conflict of interest regards data collection, processing, and interpretation in this research.

Appendix A

See **Tables A1** and **A2**.

Material	Resistivity (Ωm)	Conductivity $(\Omega m)^{-1}$
Igneous and metamorphic rocks		
Granite	5× 103–106	10–6 - 2 × 10^{-4}
Basalt	103–106	10–6 – 10^{-3}
Slate	6× 102–4 × 107	2.5 × 10^{-8} 1.7 × 10^{-3}
Marble	102–2.5 × 108	4 × 10–9 – 10^{-2}
Quartzite	102–2 × 108	5 × 10–9 – 10^{-2}
Sedimentary Rocks		
Sandstone	8–4 × 103	2.5 × 104 - 0.125
Shale	20–2 × 103	4 × 10^{-4} - 0.05
Limestone	50–4 × 102	2.5 × 10^{-3} - 0.02
Soils and Waters	1–100	1
Clay	10–800	1. 25 × 10^{-3}–0.1
Alluvium	10–100	0.01–0.1
Groundwater (fresh)	0–15	6.7
Sea water		
Chemicals and Metals		
Iron	9.074 × 10^{-8}	1.102 × 107
0.01 M Potassium chloride	0.708	1.413
0.01 M Sodium chloride	0.843	1.185
0.01 M acetic acid	6.130	0.163
Xylene	6.998 × 1016	1. 429 × 10–17

Table A1.
Resistivity of some common rocks, minerals, chemicals, and metals according to Loke [18].

Material	Resistivity (Ωm)
Common Rocks	
Top soil	50–100
Loose sand	500–5000
Gravel	100–600
Clay	1–100
Weathered bed rock	100–1000
Sandstone	200–8000
Limestone	500–10,000
Greenstone	500–200,000
Gabbro	100–500,000
Granite	200–100,000
Basalt	200–100,000
Graphitic schist	10–500
Slates	500–500,000
Quartzite	500–800,000
Ore Minerals	0.01–100
Pyrite (ores)	0.001–0.01
Pyrrhotite	0.005–0.1
Chalcopyrite	0.001–100
Sphalerite	1000–1,000,000
Magnetite	0.01–1000
Hematite	0.01–1,000,000

Table A2.
Resistivities of common rocks and mineral according to Milsom [16].

Appendix B

See **Table B1**.

No.	Location	Lat	Long
1	Rumeila 1	14.90802	35.87097
2	Rumeila 2	14.89187	35.86552
3	Munbaa 1	14.86227	35.86700
4	Munbaa 2	14.8718	35.83107
5	Munbaa	14.86216	35.86711
6	Shaqarabgreb 1	14.79306	35.88382
7	Shaqarabgreb 2	14.81162	35.87579
8	Shaqarabgreb 3	14.82960	35.81837
9	Shaqarabgreb 4	14.76759	35.86318
10	Mayela 1	14.53453	35.42737
11	Mayela 2	14.63005	35.41902
12	Mayela 3	14.64810	35.47856
13	Mayela 4	14.66880	35.54080
14	Mayela 5	14.48299	35.44783

No.	Location	Lat	Long
15	Wad Hassan 1	14.56554	35.86705
16	Wad Hassan 2	14.50723	35.77684
17	Wad Hassan 3	14.54830	35.87562
18	Magat W. Kheder	14.57200	35.88119
19	MagatA.Elnabe 1	14.68770	35.80454
20	MagatA.Elnabe 2	14.70873	35.88312
21	MagatA.Elnabe 3	14.66609	35.75485
22	MoqattaElsoog	14.64856	35.79379
23	Hagar Abyad 1	14.49048	35.87344
24	Hagar Abyad 2	14.51306	35.85472
25	Hagar Abyad 3	14.46840	35.82385
26	Hagar Abyad 4	14.47074	35.85973
27	Eldeweh1	14.77347	35.80735
28	Eldeweh2	14.73118	35.78667
29	Eldeweh3	14.83048	35.87216
30	Shengryarah 1	14.46573	35.89446
31	Shengryarah 2	14.46494	35.87563
32	Shengryarah 3	14.44529	35.86439
33	wad Elhade 1	14.54728	35.81779
34	wad Elhade 2	14.44105	35.70287
35	wad Elhade 3	14.54753	35.85257
36	MoqattaJubarab 1	14.63888	35.84559
37	MoqattaJubarab 2	14.59304	35.72556
38	MoqattaJubarab 3	14.63690	35.81965
39	MoqattaJubarab 4	14.56325	35.74353
40	ElmugasamahEl.Elnoor 1	14.61625	35.83658
41	ElmugasamahEl.Elnoor 2	14.58872	35.83129
42	ElmugasamahEl.Elnoor 3	14.61023	35.86881
43	ElmugasamahEl.Elnoor 4	14.61690	35.80793
44	Elmansurah5	14.58966	35.87083
45	ElmugasamahEl.Elnoor 6	14.61360	35.77805
46	Elmugasamah El. Elnoor 7	14.63352	35.76179
47	ElmugasamahEl.Elnoor 8	14.52974	35.73298
48	ElmugasamahEl.Elnoor 9	14.46898	35.74585
49	ElmugasamahEl.Elnoor 10	14.58772	35.80173
50	ElmugasamahEl.Elnoor 11	14.57708	35.77606
51	ElmugasamahEl.Elnoor 12	14.59512	35.88487
52	ElmugasamahEl.Elnoor 13	14.49224	35.70509
53	ElmugasamahEl.Elnoor 14	14.62504	35.88188
54	Elmugasamah El.Elnoor15	14.67594	35.87048
55	Jubarab 16	14.64507	35.87265

No.	Location	Lat	Long
56	Elmansurah 1	14.53774	35.77513
57	Elmansurah 2	14.62117	35.85451
58	Elmansurah 3	14.58422	35.85352
59	Merrebiah 1	14.43295	35.88083
60	Merrebiah 2	14.44661	35.84492
61	Merrebiah 3	14.40964	35.86676
62	Merrebiah 4	14.49423	35.83192
63	Mabroukah 1	14.52757	35.83616
64	Mabroukah 2	14.49238	35.85227
65	Mabroukah 3	14.48956	35.79208
66	Umm bade 1	14.47086	35.80158
67	Umm bade 2	14.39917	35.81764
68	Umm bade 3	14.44596	35.76924
69	Umm bade 4	14.40177	35.83800
70	Umm Gara 1	14.34408	35.78360
71	Umm Gara 2	14.35388	35.83817
72	Umm Gara 3	14.34822	35.81285
73	Um Grgour 1	14.53761	35.64524
74	Um Grgour 2	14.56020	35.51243
75	Um Grgour 3	14.56617	35.63500
76	Um Grgour 4	14.50485	35.65538
77	Um Grgour 5	14.59692	35.62541
78	Um Grgour 6	14.50426	35.52131
79	Moharagat 1	14.42204	35.73805
80	Moharagat 2	14.36743	35.73603
81	Moharagat 3	14.38601	35.69267
82	W. Alamarah 1	14.26614	35.87652
83	W. Alamarah 2	14.28993	35.85842
84	W. Alamarah 3	14.23194	35.81697
85	W. Alamarah 4	14.24404	35.86940
86	W. Alamarah 5	14.23696	35.84292
87	W. Alamarah 6	14.27796	35.83676
88	W. Alamarah 7	14.25324	35.82819
89	Abu raed 8	14.25180	35.88889
90	W. Alamarah 9	14.26273	35.80607
91	W. Alamarah 10	14.25908	35.85276
92	W. Alamarah 11	14.30276	35.84110
93	Atbara 1	14.21840	35.83704
94	Atbara 2	14.23270	35.88695

No.	Location	Lat	Long
95	Atbara 3	14.23002	35.93760
96	Atbara 4	14.20869	35.94884
97	Atbara 5	14.22455	35.89946
98	Atbara 6	14.21019	35.88609
99	Atbara 7	14.18941	35.95889
100	Atbara 8	14.17260	35.88573
101	Atbara 9	14.16474	35.84939
102	Atbara 10	14.18222	35.86330
103	Atbara 11	14.18886	35.83838
104	Atbara 12	14.22553	35.86530
105	Atbara 13	14.16225	35.97229
106	Atbara 14	14.20253	35.85710
107	Atbara 15	14.20474	35.81955
108	Elmashlab 1	14.91440	35.92534
109	Elmashlab 2	14.87481	35.92545
110	Alalem Shamal1	14.87316	35.96554
111	Alalem Shamal2	14.89551	35.92529
112	Alalem Shamal3	14.85318	35.92470
113	Alalemwasat 1	14.84983	35.94361
114	Alalemwasat 2	14.85709	36.00614
115	Alalemwasat 3	14.84302	35.98146
116	Alalemjnoob 1	14.88733	36.00228
117	Alalemjnoob 2	14.86363	35.99025
118	Alalemjnoob 3	14.87460	36.02866
119	Shaoatsherg 1	14.60569	35.93164
120	Shaoatsherg 2	14.68189	36.03224
121	Shaoatsherg 3	14.58453	35.94225
122	shaoatsherg 4	14.55208	35.90982
123	shaoatsherg 5	14.71417	36.03340
124	shaoatsherg 6	14.74725	36.03007
125	katutsherg 1	14.51362	35.88908
126	katutsherg 2	14.52882	35.90730
127	ShaqarabRuhal 1	14.74570	35.94113
128	ShaqarabRuhal 2	14.78486	36.02920
129	Shaqarabsherg 1	14.81315	36.02935
130	Shaqarabsherg 2	14.75748	35.97235
131	Shaqarabgooz 1	14.84538	36.02984
132	Shaqarabgooz 2	14.81372	36.00122
133	Shaqarabgooz 3	14.81436	35.97337

No.	Location	Lat	Long
134	Shaqarabgooz 4	14.78370	35.97892
135	Umm oud 1	14.58261	35.90577
136	Umm oud 2	14.61318	35.90329
137	Umm oud 3	14.65084	35.90065
138	Aleyan 1	14.49202	35.90748
139	Aleyan 2	14.48833	35.94333
140	Abuda 1	14.37343	35.91272
141	Abuda 2	14.36682	35.89472
142	Mellagah 1	14.33472	35.92216
143	Mellagah 2	14.32887	35.89865
144	Annater 1	14.38147	35.97167
145	Annater 2	14.43503	36.01761
146	Annater 3	14.38860	36.01954
147	Aradeiba 1	14.34174	35.86164
148	AradeibaKananah 2	14.34834	35.89486
149	Aradeiba(Hag Ahmed) 3	14.38555	35.87977
150	Aradeiba (Abuda) 4	14.36133	35.93940
151	Aradeiba(Kananah) 5	14.40236	35.89529
152	Aradeiba(Hag Ahmed) 6	14.36221	35.86814
153	Setit 1	14.28336	36.04842
154	Setit 2	14.29044	35.99614
155	Setit 3	14.22741	35.95343
156	Setit 4	14.28382	35.95929
157	Setit 5	14.26435	35.95479
158	Setit 6	14.32898	36.04676
159	Setit 7	14.34638	35.96806
160	Setit 8	14.31487	36.00499
161	Setit 9	14.32897	36.01768
162	Setit 10	14.30118	35.95021
163	Setit 11	14.29188	35.91301
164	Setit 12	14.17616	36.02429
165	Setit 13	14.27555	35.93444
166	Setit 14	14.30862	35.97997
167	Setit 15	14.33063	35.99184
168	Setit 16	14.32931	35.94728
169	Setit 17	14.20636	35.98352
170	Setit 18	14.27182	35.98480
171	Setit 19	14.26229	36.01323
172	Setit 20	14.31131	36.03410
173	Setit 21	14.25328	35.98850

No.	Location	Lat	Long
174	Setit 22	14.23787	36.01084
175	Wad Elhelew 1	14.24688	36.04939
176	Wad Elhelew 2	14.26353	36.03409
177	Wad Elhelew 3	14.28566	36.02401

Table B1.
The number and names of the VES points carried out in the study area.

Appendix C

See **Table C1**.

	Borehole name	Lat	Long	Elevation (m)
1	Rumeila(Alryan)	14.90113	35.86875	480
2	Rumeila	14.90356	35.85977	480
3	Rumeila—S	14.88038	35.86427	470
4	Munbaa1	14.86172	35.86605	481
5	Eldeweh	14.83742	35.88156	497
6	Shaqarab greb2	14.81408	35.90213	477
7	Wad Ebuied	14.75406	35.8949	481
8	Wad Essiea	14.74356	35.89322	496
9	Wad Hassan3	14.57081	35.88	500
10	MoqattaJubarab	14.63607	35.87228	499
11	Moqatta Wad Elzein	14.6469	35.8654	489
12	MoqattaElsoog	14.66849	35.8654	493
13	ElmugasamahelshekhElnoor	14.5994	35.86939	493
14	Almunsora	14.5919	35.86667	498
15	AlmadinahAarab	14.59172	35.8667	500
16	Magate	14.57908	35.87306	496
17	Magate shoot Shamal	14.5701	35.88788	496
18	Magate Wad Mohamadeen	14.56448	35.88847	504
19	MagatAbd Elnabe2	14.69407	35.8726	489
20	AlshareefHasaballa	14.557	35.87641	492
21	Wad Elhade	14.53368	35.85912	503
22	Mabroukah—New	14.50885	35.85252	509
23	Hagar Abyad	14.47384	35.88191	498
24	Shengryarah	14.4594	35.88012	494
25	Merrebiah	14.44044	35.88	491
26	Umm Gargoor 1	14.50121	35.62597	538
27	Umm Gargoor 2	14.50774	35.63789	532

	Borehole name	Lat	Long	Elevation (m)
28	Karkora1	14.50532	35.62244	517
29	Showak10	14.37022	35.81721	510
30	Showak20	14.38976	35.82773	509
31	Showak25	14.4073	35.83725	508
32	Showak6	14.42383	35.84527	507
33	Showak5	14.3722	35.83463	503
34	Showak31	14.39126	35.8573	502
35	Showak2	14.41076	35.87641	501
36	Alamara Wad Elzein	14.42849	35.86811	516
37	El Muneira	14 13.933	35 53.084	514
38	Tommat Wad Zaied	14.29857	35.84453	489
39	Abu read	14 15.780	35 53.372	516
40	Albushra	14.2585	35.82645	526
41	AlalemShamal	14.8651	35.91571	478
42	Alalemwasat	14.83867	35.92888	485
43	Alalemjnoob	14.49.904	35 56.538	482
44	Shaoatsherg 1	14.57535	35.91476	486
45	Shaoat sherg2	14.57245	35.91738	490
46	Almugatahshreg	14.67201	35.89497	484
47	Umm oud	14.59826	35.91211	501
48	AlshareefHammad	14.54845	35.91519	506
49	Alkarada	14.53483	35.894	502
50	Elyan	14 30.478	35 53.156	498
51	Elyan village	14.50737	35.88513	498
52	Abbuda	14.35693	35.89009	510
53	Abbuda2	14.35883	35.88481	510
54	Abbuda 3	14.36109	35.88775	510
55	Abbudda Sudanese	14.351	35.88104	520
56	Umm Ali	814,793	1,581,953	510
57	Mellagah1	14.31085	35.91122	514
58	Zakria1- East	14.28407	35.93329	516
59	Zakria2- West	14.28345	35.93063	514
60	Wad Elhelew 1	14.27132	36.00191	524
61	Wad Elhelew 2	14.24796	36.03106	529
62	WadElhelew3	14.22031	36.05302	522
63	WadElhelew4	14.22553	36.02992	521
64	WadElhelew5	14.25208	36.0004	519

Table C1.
The number and coordinates of the monitored boreholes in the study area.

DOI: http://dx.doi.org/10.5772/intechopen.1007086

Author details

Khalid Nayl[1*], Abdalla Elsheikh[2], Adil Elkrail[2] and Abobaker Elbahari[3]

1 Groundwater and Wadies Directorate, Ministry of Water Resources, Kassala, Sudan

2 Department of Hydrogeology, Faculty of Petroleum and Minerals, Al Neelain University, Khartoum, Sudan

3 Arab Consult for Geology and Environment, Riyadh, Saudi Arabia

*Address all correspondence to: nail62964@yahoo.com

References

[1] El-Galladi AM, Mohamed A, Sultan A. Mapping peat layer using surface geo-electrical method at Mansoura environs, Nile Delta, Egypt. Mansoura Journal of Geology and Geophysics. 2007;**34**(1):50-69

[2] Koefoed O. Geosounding Principles 1: Resistivity Sounding Measurements. Amsterdam: Elsevier Science Publishing Company; 1979

[3] Keller GV, Frischknecht FC. Electrical Methods in Geophysical Prospecting. Oxford: Pergamon Press Inc.; 1966

[4] Telford WM, Geldart LP, Sheriff RE. Applied Geophysics. 2nd ed. New York, USA: Cambridge University; 1990

[5] Satpathy BM, Kanungo DN. Groundwater exploration in hard rock terrain. A Case History, Geophysical Prospecting. 1976;**24**:725-736

[6] Mbonu PDC, Ebeniro JO, Ofoegbu CO, Ekine AS. Geo-electric sounding for the determination of aquifer characteristics in parts of the Unuahia area of Nigeria. Journal of Geophysics. 1991;**56**(2):284-291

[7] Choudhury K, and. Saha, D. K. Integrated geophysical and chemical study of saline water ntrusion: NGWA. Ground Water. 2004;**42**(5):671-677

[8] Kelly WE. Geo-electrical sounding for estimating aquifer hydraulic conductivity. Ground Water. 1977;**15**:420-424

[9] Okereke CS, Esu EO, Edet AE. Determination of potential groundwater sites using geological and geophysical techniques in the Cross River state, South-Eastern Nigeria. Journal of African Earth Sciences. 1998;**27**(1):149-163

[10] Hago A, Hago. Application of electrical resistivity method in quantitative assessment of groundwater reserve of unconfined aquifer, Bukit Jalil, Selangor D.E. Area, Malaysia [M.Sc. thesis]. Malaysia: Universiti Putra Malaysia Institutional Repository; 2000. 194 pp

[11] Telford WM, Geldart LP, Sheriff RE, Keys DA. Applied Geophysics. 2nd ed. Cambridge: Cambridge University Press; 1982

[12] Huntley D. Relations between permeability and electrical resistivity in granular aquifers. Groundwater. 1986;**24**:466-474

[13] Mazac O, Kelly WE, Landa I. A hydro-geophysical model for relations between electrical and hydraulic properties of aquifers. Journal of Hydrology. 1985;**79**:1-19

[14] Onuoha KM, Mbazi FCC. Aquifer transmissivity from electrical sounding data: The case of Ajali sandstone aquifers south-west of Enugu, Nigeria. In: Ofoegbu CO, editor. Groundwater and Mineral Resources of Nigeria. Vieweg - Verlag. Glendale CA, USA: Scientific Research Publishing (SCIRP); 1988. pp. 17-30

[15] Niwas S, Singhal DC. Estimation of aquifer transmissivity from Dar Zarrouk parameters in porous media. Hydrology. 1981;**50**:393-399

[16] Milsom J. Field Geophysics (the Geological Field Guide Series). 3rd ed. England: John Wiley and Sons Ltd; 2003. 493 pp

[17] Zohdy AAR, Eaton GP, Mabey DR. Application of Surface Geophysics to Groundwater Investigations. Techniques of Water Resources Investigations of the United States-Geological Survey. USA: U.S. Geological Survey; 1974

[18] Loke MH. Electrical Imaging Survey for Environmental and Engineering Studies. Unpublished Report; 1997

Chapter 5

Integrated Geophysical Techniques for Subsurface Characterization and Groundwater Assessment: A Case Study from the University of Benin, Nigeria

Efetobore Gladys Maju-Oyovwikowhe, Ekenakema Uwa-Igbinoba and Owens Monday Alile

Abstract

This chapter investigates the application of integrated geophysical surveys, specifically electrical resistivity and seismic refraction methods, to characterize the subsurface at the University of Benin, Nigeria. The study's primary objective was to delineate subsurface geological features, evaluate groundwater potential, and assess environmental conditions critical for sustainable resource management and infrastructure development. The integration of electrical resistivity and seismic refraction techniques provided a comprehensive understanding of the subsurface, revealing significant variations in geological structures, identifying high-yield aquifer zones, and evaluating geotechnical properties essential for construction and environmental planning. The chapter explores the effectiveness of these integrated methods in complex geological settings, addressing the challenges encountered and proposing strategies for optimizing data acquisition and interpretation. The findings highlight the importance of employing multiple geophysical approaches to obtain precise and reliable subsurface models, which are vital for informed decision-making in groundwater management and environmental stewardship. This case study offers valuable insights and practical guidelines for the application of geophysical methods in similar geological contexts, aiming to enhance the sustainability of groundwater resources and improve the resilience of environmental systems. Through these efforts, the study contributes to the advancement of environmental geophysics and its applications in resource management.

Keywords: geophysical surveys, electrical resistivity, seismic refraction, subsurface characterization, groundwater assessment, environmental geophysics, University of Benin

1. Introduction

Geophysical surveys are essential tools in understanding the Earth's subsurface, providing vital information that informs a variety of applications such as groundwater exploration, environmental assessments, and infrastructure development. These surveys employ a range of techniques, with electrical resistivity tomography (ERT) and seismic refraction emerging as particularly effective methods for subsurface characterization, especially in regions with complex geological settings. This chapter explores the application of these integrated geophysical techniques at the University of Benin, Nigeria, aiming to delineate subsurface geological features, evaluate groundwater potential, and assess environmental conditions.

1.1 Context and importance of geophysical surveys

Geophysical methods have long been recognized for their ability to provide non-invasive, detailed insights into the subsurface. ERT, for instance, measures the resistivity distribution of subsurface materials, making it particularly effective in identifying variations in lithology, fluid content, and porosity. This technique is invaluable in groundwater exploration and environmental studies where understanding the distribution of these properties is crucial for effective resource management [1, 2]. Seismic refraction, on the other hand, measures the velocity of seismic waves as they travel through different subsurface layers. This method is highly effective in delineating layer boundaries, determining depth to bedrock, and assessing the mechanical properties of subsurface materials [3, 4]. The integration of these methods has proven successful in a variety of contexts, enhancing the accuracy of subsurface models and reducing the risk of misinterpretation [2, 5]. Recent studies have highlighted the growing importance of multi-dimensional geophysical techniques in environmental studies, particularly in agrogeophysics, where ERT has been increasingly used to monitor soil moisture dynamics and subsurface properties [6].

1.2 Application at the University of Benin

The University of Benin, located in a geologically diverse region, serves as an ideal study area for applying these integrated geophysical methods. The location map of the study area, as shown in **Figure 1**, highlights the specific sites where the vertical electrical sounding (VES) and seismic spreads were conducted across the University of Benin campus. This study builds on previous research that underscores the importance of integrating multiple geophysical techniques to achieve more reliable and accurate subsurface characterizations [7, 8]. By applying ERT and seismic refraction in tandem, this research aims to provide a detailed understanding of the subsurface conditions, identifying critical features such as aquifer zones, fault lines, and varying soil compositions.

1.3 Relevance and implications

The findings from this study have broad implications, not only for the University of Benin but also for other regions with similar geological conditions. Effective subsurface characterization is crucial for sustainable groundwater management, environmental protection, and infrastructure development [9, 10]. By integrating geophysical methods, this research contributes to the development of best practices

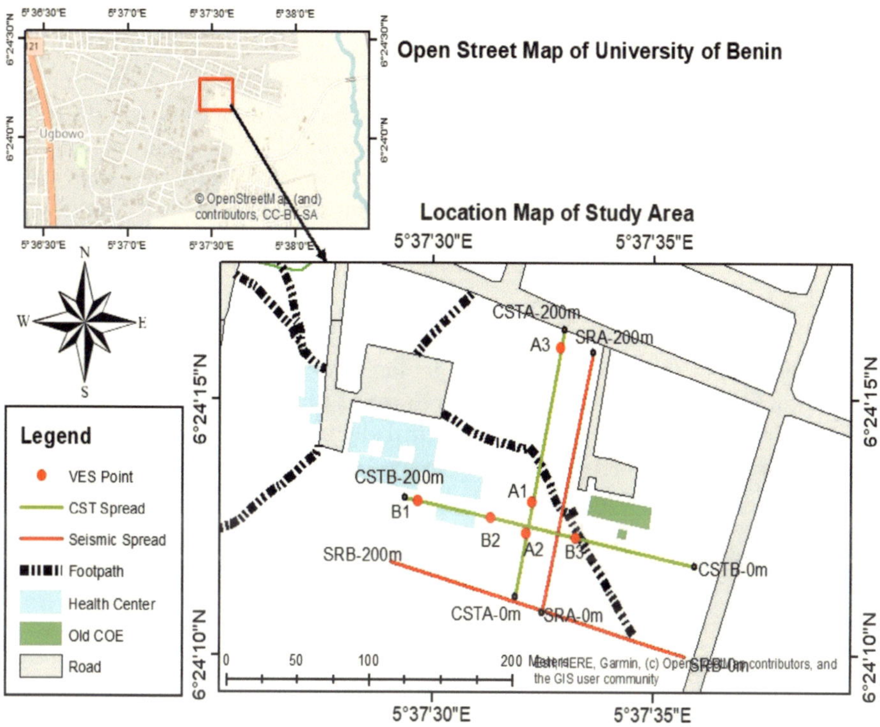

Figure 1.
Location map of the study area.

in subsurface exploration, promoting the use of these techniques in diverse settings to enhance the accuracy and reliability of subsurface investigations.

In conclusion, this chapter provides a comprehensive overview of the application of integrated geophysical methods for subsurface characterization at the University of Benin. The insights gained from this study are intended to guide similar applications in other regions, contributing to the broader field of geophysical research and its applications in resource management and environmental stewardship.

2. Methodology and data analysis

This section outlines the comprehensive methodology employed in this study, focusing on the integration of electrical resistivity tomography (ERT) and seismic refraction techniques to investigate the subsurface characteristics at the University of Benin, Nigeria. The section details the data acquisition processes, the equipment used, and the subsequent data processing and interpretation techniques applied. The integration of these geophysical methods provides a robust framework for understanding the subsurface conditions, which is crucial for environmental and geotechnical applications. The integration of ERT and seismic refraction has proven to be highly effective in subsurface investigations, as demonstrated in recent studies focusing on sinkhole detection in karst areas [11]. The application of integrated near-surface geophysical techniques, including ERT and seismic refraction, has been shown to improve subsurface characterization in complex terrains, further validating the methodologies employed in this study [12].

2.1 Electrical resistivity tomography (ERT)

ERT is a geophysical method used to map subsurface resistivity by injecting electrical current into the ground and measuring the resulting potential differences. The resistivity of the subsurface is influenced by various factors, including the type of soil or rock, moisture content, and the presence of voids or fractures. This method is particularly useful for identifying variations in subsurface lithology, detecting groundwater, and mapping contaminant plumes [1, 2].

2.1.1 Study area

The University of Benin is situated in a geologically diverse region, making it an ideal location for subsurface investigations. **Figure 1** illustrates the study area, highlighting the specific locations where the vertical electrical sounding (VES) points and seismic spreads were conducted. The map shows the distribution of the survey points, with details on the seismic and CST spreads.

2.1.2 Geological context

Understanding the geological context is vital for interpreting the results of geophysical surveys. **Figure 2** presents the geological map of Nigeria, as produced by the

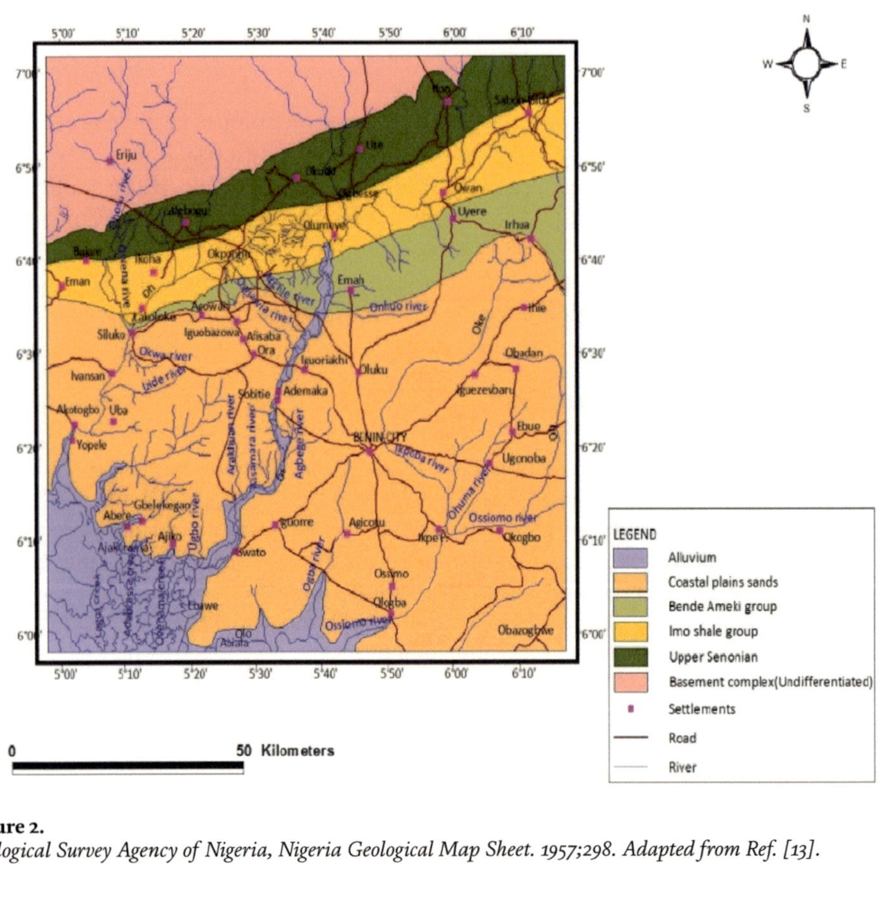

Figure 2.
Geological Survey Agency of Nigeria, Nigeria Geological Map Sheet. 1957;298. Adapted from Ref. [13].

Geological Survey Agency of Nigeria in 1957 and later adapted by Olatunji et al. [13]. This map provides an overview of the regional geology, which underpins the subsurface characteristics examined in this study. The map highlights key geological formations and structures that are relevant to the study area, particularly the lithological units that were likely encountered during the surveys.

2.1.3 Data acquisition

For this study, data were collected using an ABEM Terrameter LS 2. The survey was conducted using a Wenner array configuration, known for its sensitivity to both lateral and vertical variations in subsurface resistivity. The field configuration for the Wenner array is illustrated in **Figure 3**, providing a clear layout of the electrode placements used in the study. Two profiles were surveyed, each 200 meters in length, with electrodes spaced at 5-meter intervals. The ERT data were captured along multiple profiles to ensure comprehensive coverage of the study area [2, 14, 15].

Additionally, the Schlumberger array configuration was also employed, as shown in **Figure 4**. This configuration differs from the Wenner array by having varying

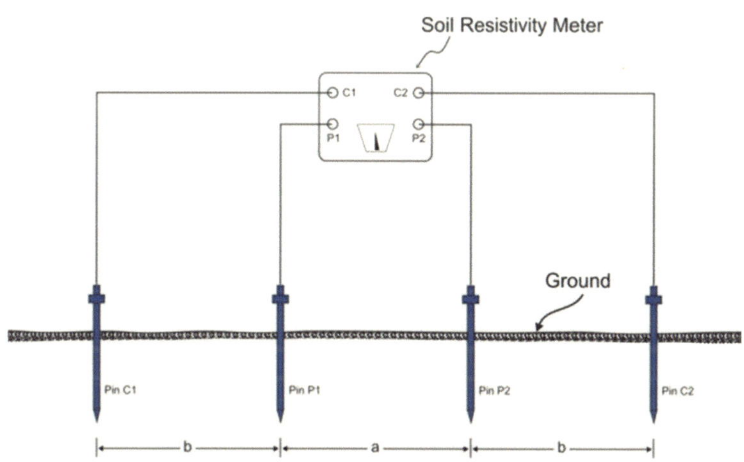

Figure 3.
Field configuration for Wenner array.

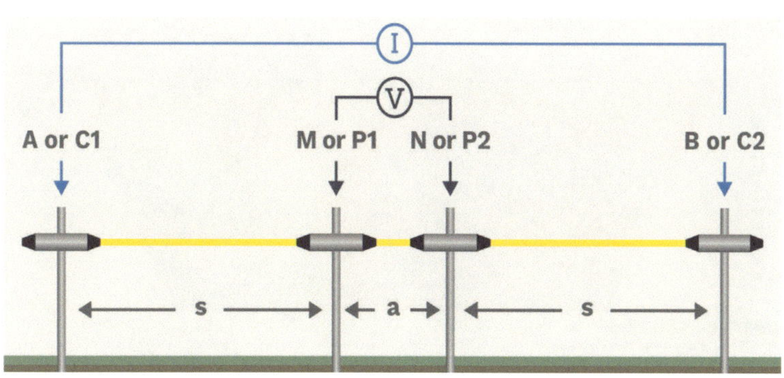

Figure 4.
Field configuration for Schlumberger array.

electrode spacing, which provides different depth penetration and is particularly useful for deeper subsurface investigations.

3. Results and discussion

3.1 Electrical resistivity tomography (2D survey)

The results from the electrical resistivity tomography (ERT) data collected along two transects are presented in **Figures 5** and **6**. These figures display contoured apparent resistivity pseudo-sections that provide insights into the subsurface resistivity distribution within the study area. The 2D resistivity structure sections, derived from the inversion of these pseudo-sections, show computed resistivity values with depth, following the methodology outlined by Alile et al. [15].

The 2D resistivity images reveal the subsurface resistivity characteristics at various depths. Generally, the depth of investigation is proportional to the separation between electrodes [16]. The resistivity profiles suggest the presence of four distinct geoelectrical zones, which are inferred to represent topsoil, sand, dry sand, and

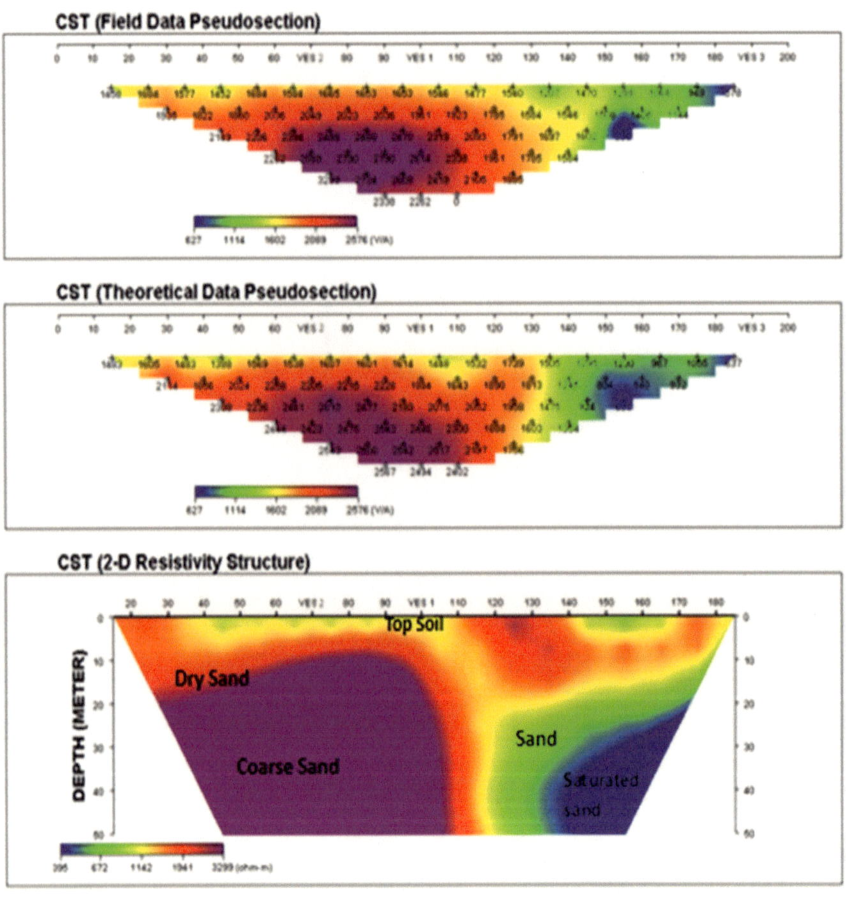

Figure 5.
Subsurface image for profile 1.

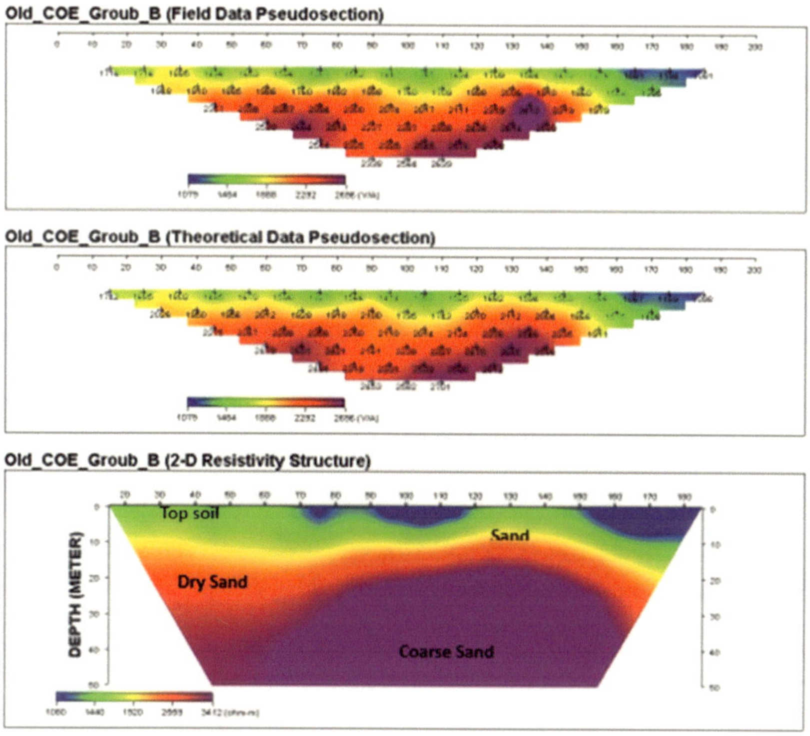

Figure 6.
Subsurface image for profile 2.

saturated sand. These inferences are based on resistivity ranges previously reported by Egbo and Airen [7].

3.1.1 Data acquisition

For this study, data were collected using an ABEM Terrameter LS 2. The survey was conducted using a Wenner array configuration, known for its sensitivity to both lateral and vertical variations in subsurface resistivity. The field configuration for the Wenner array is illustrated in **Figure 3**, providing a clear layout of the electrode placements used in the study. Two profiles were surveyed, each 200 meters in length, with electrodes spaced at 5-meter intervals. The ERT data were captured along multiple profiles to ensure comprehensive coverage of the study area [2, 14]. Additionally, the Schlumberger array configuration was also employed, as shown in **Figure 4**. This configuration differs from the Wenner array by having varying electrode spacing, which provides different depth penetration and is particularly useful for deeper subsurface investigations.

3.1.2 Data processing

The acquired ERT data were processed using the RES2DINV software, which employs an inversion algorithm to convert the apparent resistivity data into a true resistivity model. This process involved several steps, including data filtering, inversion, and model refinement. The resulting resistivity sections were then interpreted to identify key subsurface features such as aquifer zones, fault lines, and variations in soil composition. **Figure 5** shows a representative subsurface image from Profile 1, where distinct resistivity contrasts were observed, indicating the presence of multiple geoelectrical layers [14].

Profile 1 (**Figure 5**) reveals a highly productive zone for groundwater extraction, particularly in the lower eastern part of the profile where a saturated sand section is evident. This suspected high-potential zone was further investigated using vertical electrical sounding (VES 3), which will be discussed in detail in the following sections. To enhance the comprehensive understanding, two other suspected productive zones were also probed using VES 1 and VES 2.

Conversely, Profile 2 (**Figure 6**) appears to be less productive for groundwater, as it lacks a clear saturated sand lithology. However, this does not necessarily indicate a non-water-bearing area. The zones with suspected groundwater potential were further investigated using VES 4, VES 5, and VES 6, and these findings will also be detailed in the subsequent sections.

The average apparent resistivity across the study area is approximately 1500 Ωm, which is typical for sedimentary environments and aligns with the general geology of the region. Wet soils and newly formed groundwater generally exhibit lower resistivity values compared to drier soils and sediments [9].

Ogungbe et al. [9] highlight that road failures and engineering challenges are often linked to poor subgrade materials and water-logged sands, leading to major cracks and structural failures. The presence of clay soils exacerbates these issues due to their ability to absorb water, swell, and fail under stress, such as heavy vehicle movement or construction activities. Fortunately, the study area appears to be suitable for road and building construction, as the subsurface images do not indicate significant clay layers.

3.2 Vertical electrical sounding (VES)

Six VES surveys were conducted in zones with anomalous resistivity identified in the ERT imaging. The VES data were quantitatively interpreted using partial curve matching and computer-assisted iterative techniques with the WINRESIST software. The computer modeling used the initial layer resistivities and thicknesses as input for the geological interpretation, which was further informed by the previously generated ERT subsurface sections.

The geoelectric parameters obtained from the VES surveys are summarized in **Table 1**, which shows the resistivity, thickness, depth, curve type, and inferred lithology for each VES point. The table reveals distinct lithological layers consistent with previous studies in the region. Alile et al. [15] identified similar resistivity values and lithologies in their study of the Benin formation, reinforcing the accuracy of the interpretations provided here. For instance, the saturated sand layers crucial for groundwater potential exhibit resistivity ranges aligned with those reported [15].

The VES surveys identified significant lithological features across the study area, with varying groundwater potential ratings. These results highlight the heterogeneity of the subsurface, confirming the necessity for detailed geophysical investigations in complex geological settings.

3.3 Seismic refraction survey

The seismic refraction survey results are presented in **Figures 7–12**, which show the S-wave and P-wave velocity models for Profiles 1 and 2. These figures reveal variations in subsurface velocities, which are crucial for understanding the geotechnical properties of the study area. The calculated elastic moduli and geotechnical parameters, based on the seismic data, are summarized in **Table 2**. These parameters

VES NO	Layers	Resistivity	Thickness(m)	Depth(m)	Curve type	Inferred lithology [7]	Groundwater potential rating
VES 1	1	371.5	0.8	0.8	KHK	Top soil	Medium
	2	2603.9	3.6	4.4		Dry sand	
	3	2416.1	7.0	11.4		Dry sand	
	4	3375.6	47.5	58.9		Dry sand	
	5	869.8				Saturated sand	
VES 2	1	367	1	1	AKH	Top soil	Medium
	2	2090.9	4	5		Dry sand	
	3	3132.8	5.9	10.9		Dry sand	
	4	1083.2	14.5	25.5		Saturated sand	
	5	6064.5				Coarse sand	
VES 3	1	246.1	1.1	1.1		Top soil	High
	2	312.2	2.3	3.4	AKH	Sand	
	3	742.3	8	11.4		Saturated sand	
	4	567.1	45.3	56.8		Saturated sand	
	5	2296.2				Dry sand	
VES 4	1	840	1	1	KHA	Top soil	Low
	2	1432.5	2.1	3.1		Dry sand	
	3	1289.5	3.8	6.9		Dry sand	
	4	1856.8	15.6	22.5		Dry sand	
	5	4686.4				Dry sand	
VES 5	1	262.7	0.9	0.9	KHA	Top soil	Medium
	2	1400.7	2.2	3.1		Dry sand	
	3	839.7	8.2	11.3		Saturated sand	
	4	2222.1	13.7	25		Dry sand	
	5	15751.7				Coarse sand	
VES 6	1	564	0.8	0.8	KH	Top soil	Low
	2	1715.3	7.7	8.5		Dry sand	
	3	1193	15.6	24		Dry sand	
	4	10,124				Coarse sand	

Table 1.
Geoelectric parameters and their inferred lithology.

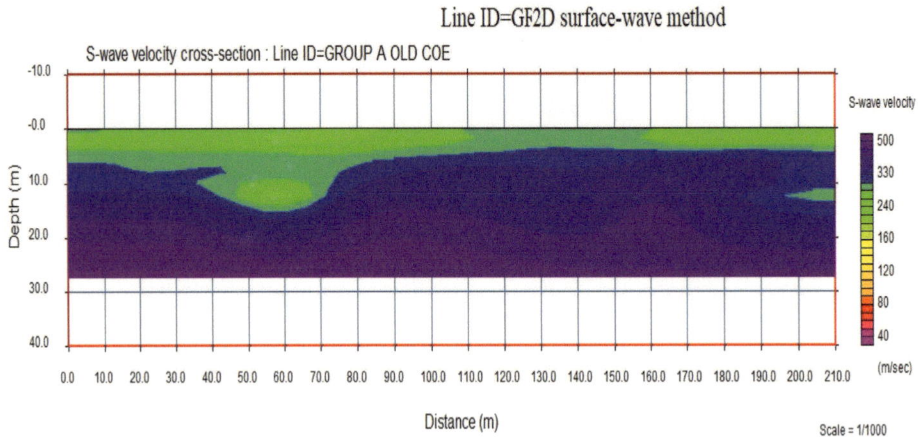

Figure 7.
Profile 1 S wave velocity model.

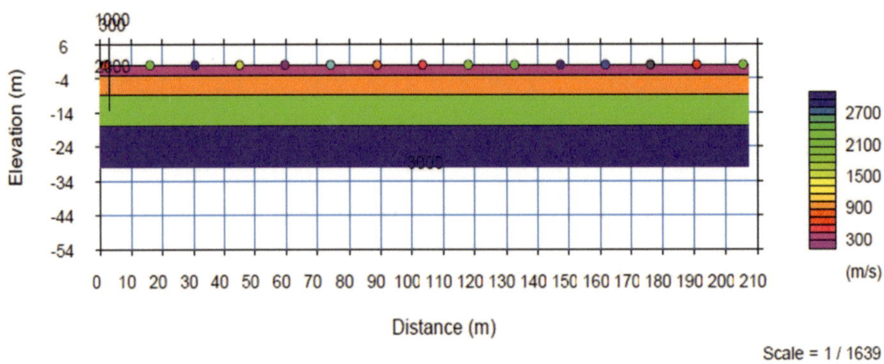

Figure 8.
Profile 1 P wave velocity model.

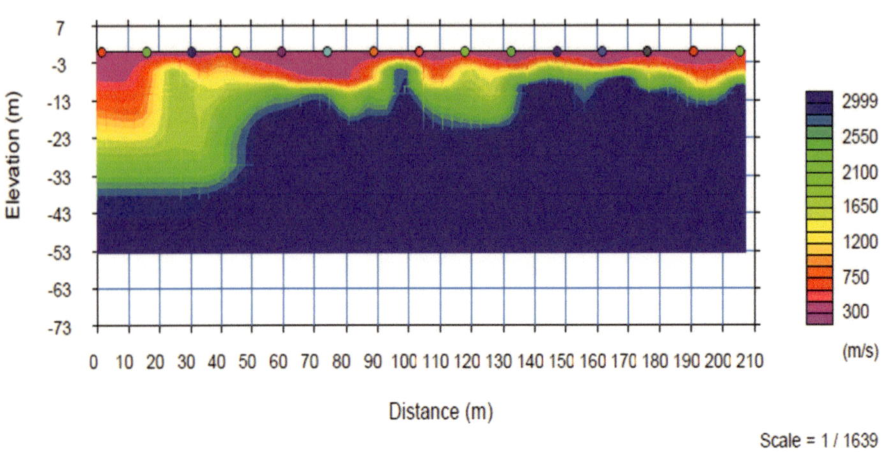

Figure 9.
Profile 1 inverted P wave velocity model.

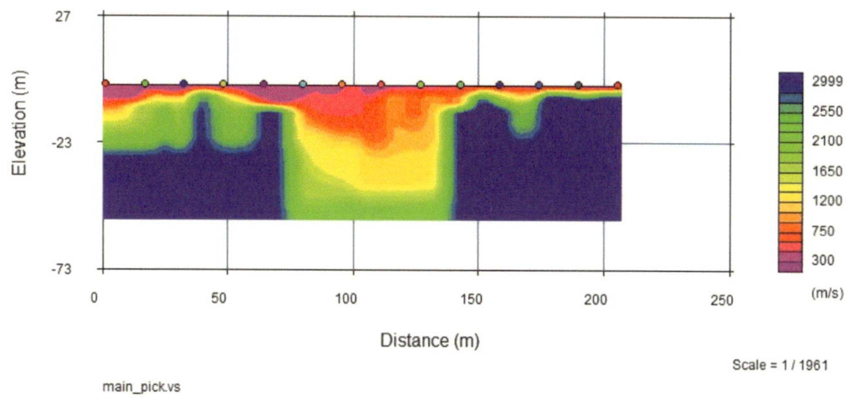

Figure 10.
Profile 2 inverted P wave velocity model.

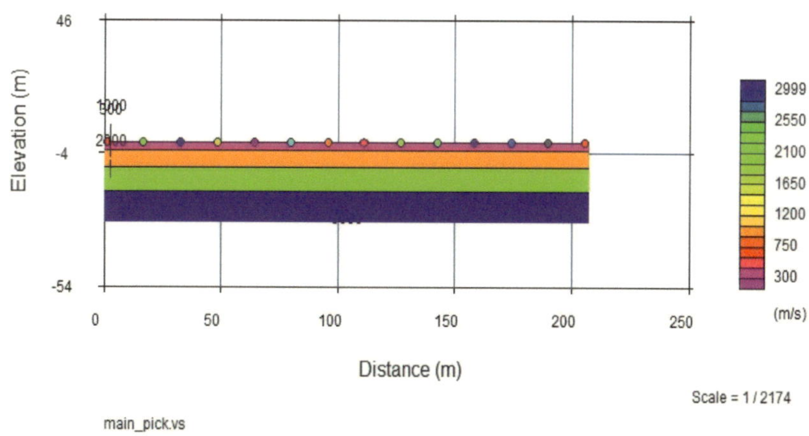

Figure 11.
Profile 2 P wave velocity model.

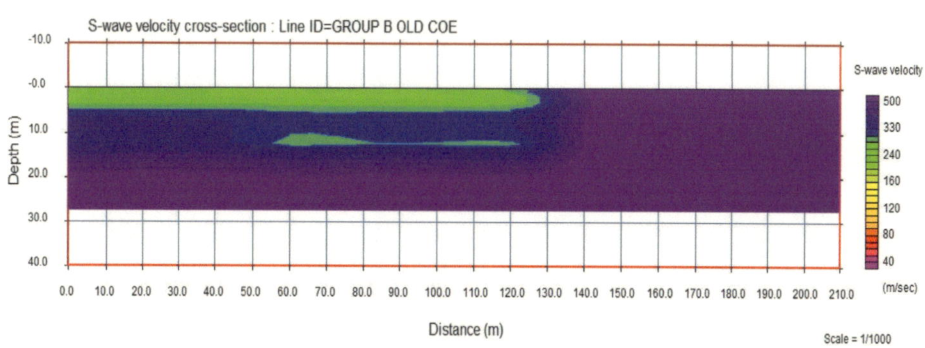

Figure 12.
Profile 2 S wave velocity model.

Vp	Vs	Poisson ratio	Density	Young's modulus	Lame's constant	Shear modulus	Concentration ratio	Material index	Stress ratio
300	240	−0.38889	1.249	0.088	−0.0315	0.0719	−1.571428571	2.556	−0.280
525	370	0.006578	1.436022	0.396	0.0026	0.1966	153.0273973	0.9737	0.00662

Table 2.
Calculated elastic moduli and geotechnical parameters.

are critical for assessing the suitability of the site for construction and other engineering purposes [3, 5].

Table 2 provides a detailed breakdown of wave velocities (Vp and Vs), Poisson's ratio, density, Young's modulus, Lame's constant, shear modulus, and other relevant geotechnical indices. By comparing these results with established engineering standards, civil engineers can make informed decisions regarding the design and safety of structures to be built on this site.

3.3.1 Data processing

The seismic data were processed using SeisImager software. The processing involved picking first-arrival times, applying static corrections, and generating velocity-depth models. The models produced from the seismic data were used to interpret the subsurface geology and to correlate with the ERT data. **Figures 10** and **11** show the S-wave and P-wave velocity models for Profile 1, respectively, which were critical in determining the geotechnical properties of the subsurface [4].

3.4 Integration and interpretation of geophysical data

The integration of electrical resistivity tomography (ERT) and seismic refraction data provides a comprehensive understanding of the subsurface conditions. By correlating the resistivity and seismic velocity models, a more accurate interpretation of the subsurface geology was achieved. This combined interpretation allows for the identification of key features such as potential aquifer zones, fault lines, and areas of varying soil stability [1, 10].

Low-resistivity zones identified in the ERT profiles correlate with low-velocity zones in the seismic data, suggesting the presence of water-saturated layers or potential aquifers. These zones are crucial for groundwater exploration and management. For instance, the high groundwater potential indicated in **Table 1** aligns with areas where these low-resistivity and low-velocity zones overlap, confirming the presence of viable aquifers [8, 16].

The seismic velocity models revealed significant differences in the subsurface material properties across the study area. For instance, **Table 2** shows that Profile 2 had higher P-wave and S-wave velocities, indicating more competent and stable subsurface materials suitable for construction. These findings suggest that certain areas are more favorable for building infrastructure due to the stability of the subsurface materials [3, 9].

The integration of ERT and seismic refraction data has demonstrated the effectiveness of using multiple geophysical methods to achieve a detailed and reliable subsurface characterization. This approach not only enhances the understanding of the subsurface geology but also provides critical insights for environmental and geotechnical applications. For instance, the correlated data from **Tables 1** and **2** offer a robust framework for identifying areas with both high groundwater potential and suitable geotechnical properties for construction [2, 5, 15]. Additionally, this methodology is supported by similar studies that have shown the value of integrating geophysical techniques to improve subsurface characterization accuracy [4, 17].

By applying these integrated methods, the study contributes to a deeper understanding of the subsurface, which is essential for informed decision-making in environmental and geotechnical projects, such as groundwater management, construction planning, and hazard assessment [1, 18, 19]. The successful application of time-lapse

ERT in monitoring permafrost dynamics in the Arctic illustrates the method's versatility and its potential application in various environmental conditions [20]. ERT has been effectively used to monitor soil and subsoil conditions, providing critical insights into environmental responses, such as tree responses to drought, which is pertinent to understanding subsurface conditions in the study area [21].

4. Conclusions

This chapter has presented a comprehensive study on the integration of electrical resistivity tomography (ERT) and seismic refraction techniques for subsurface characterization at the University of Benin, Nigeria. The research demonstrated the effectiveness of these geophysical methods in providing detailed insights into the subsurface geology, which is crucial for applications such as groundwater exploration, environmental assessments, and geotechnical evaluations.

4.1 Key findings

1. *Subsurface characterization*: The ERT and seismic refraction surveys successfully delineated subsurface features, identifying key layers such as topsoil, dry sand, and saturated sand. The correlation between resistivity and seismic velocity models allowed for a more accurate interpretation of subsurface conditions, reinforcing the importance of integrating multiple geophysical techniques for comprehensive subsurface analysis [1, 2, 5].

2. *Aquifer identification*: Low-resistivity and low-velocity zones identified potential aquifer layers, which are critical for sustainable groundwater management. These findings can guide future groundwater exploration and exploitation in the study area, contributing to more effective and sustainable water resource management practices [8, 17].

3. *Geotechnical properties*: The seismic data provided valuable information on the geotechnical properties of the subsurface materials, highlighting areas of higher stability suitable for construction and infrastructure development. This is particularly significant in regions with variable geological conditions, where understanding subsurface stability is essential for safe construction practices [3, 9].

4.2 Implications and recommendations

The successful integration of ERT and seismic refraction methods in this study underscores the importance of using multiple geophysical techniques for comprehensive subsurface investigations. Recent advancements in geophysical surveys, such as the hydrochemical and isotopic studies of coastal aquifers in Douala [22] and the integration of remote sensing with geophysical methods for groundwater potential mapping [23], highlight the growing need for multidisciplinary approaches in subsurface characterization.

Additionally, the application of these techniques in varied contexts, including the monitoring of seawater intrusion in coastal plains [10, 24], demonstrates their versatility and effectiveness in addressing complex environmental challenges. Future studies could benefit from incorporating these recent developments to further enhance the resolution and accuracy of subsurface models [18, 19, 25].

4.3 Final thoughts

In conclusion, this chapter contributes to the broader field of geophysical research by providing a detailed case study that illustrates the value of combining ERT and seismic refraction techniques. The insights gained from this study not only enhance our understanding of the subsurface geology in the University of Benin area but also provide practical guidelines for environmental management, groundwater exploration, and infrastructure planning in similar settings. These findings align with recent studies that emphasize the importance of integrating multiple geophysical methods to achieve a comprehensive understanding of subsurface conditions [1, 16, 22, 23].

Acknowledgements

The authors would like to express their deepest gratitude to the Department of Geology, University of Benin, for providing the necessary resources and support throughout the research process. We are particularly grateful to the staff and students of the Center of Excellence in Geosciences and Petroleum Engineering for their technical assistance during the fieldwork and data acquisition phases. Special thanks go to Professor Okpeseyi Isaac Imasuen for his invaluable guidance and mentorship.

Lastly, we are thankful to all the unnamed individuals whose contributions, both large and small, facilitated the completion of this work. Your support, whether through logistical help or intellectual discourse, has been indispensable, and we are deeply appreciative.

Conflict of interest

The authors declare no conflict of interest.

Notes/thanks/other declarations

The authors would like to extend special thanks to the field assistants and laboratory technicians whose diligence and hard work were crucial in the successful completion of this study.

A. Appendices

See **Tables A1–A3**.
See **Figures A1–A6**.

Elastic modulus	Used equation	References
Poisson's ratio	$\sigma = \frac{1}{2}\left(1-\left\{\frac{1}{\left(\left(\frac{v_p}{v_s}\right)^2-1\right)}\right\}\right)$	Adams and Williamson [26], Salem [27]
Young's modulus	$E = \left[\frac{\left(3V_p^2-4V_s^2\right)}{\left(\left(\frac{V_p}{V_s}\right)^2-1\right)}\right]$	Adams and Williamson [26]
Lame's constants	$\lambda = \frac{\sigma E}{((1+\sigma)(1-2\sigma))}$	King [28], Toksoz et al. [29]
Density	$ñ = \left[0.3V_p^{0.25}\right]$	Gardner et al. [30]
Shear modulus	$ì = \frac{E}{(2(1+\sigma))}$	King [28], Toksoz et al. [29]

Table A1.
Equations of elastic moduli used. V_p *and* V_s *are P-and S-waves velocities, respectively.*

Engineering parameter	Used equation	References
Concentration index	$C_i = \frac{\left[3-4\left(\frac{v_v^2}{v_D^2}\right)\right]}{\left[1-2\left(\frac{v_v^2}{v_D^2}\right)\right]}$	Abd El-Rahman [31]
Material index	$v = \frac{(\mu-\lambda)}{(\mu+\lambda)} = (1-4\sigma)$	Abd El-Rahman [32]
Stress ratio	$S_i = \left[1-2\left(\frac{V_s^2}{V_D^2}\right)\right] = (C_i-2)^{-1}$	Abd El-Rahman [31]

Table A2.
Engineering parameter equations used.

Soil description parameter	Incompetent to slightly competent	Fairly to moderately competent	Competent materials	Very high competent materials
Poisson's ratio (σ)	0.41–0.49	0.35–0.27	0.25–0.16	0.12–0.03
Material index (v)	(−0.5)–(−1)	(−0.5)–(0.0)	0.0–0.5	>0.5

Table A3.
Soil description with respect to Poisson's ratio and material index, after Birch [33], Gassman [34], Tatham [35], and Sheriff and Geldart [36].

DOI: http://dx.doi.org/10.5772/intechopen.1006853

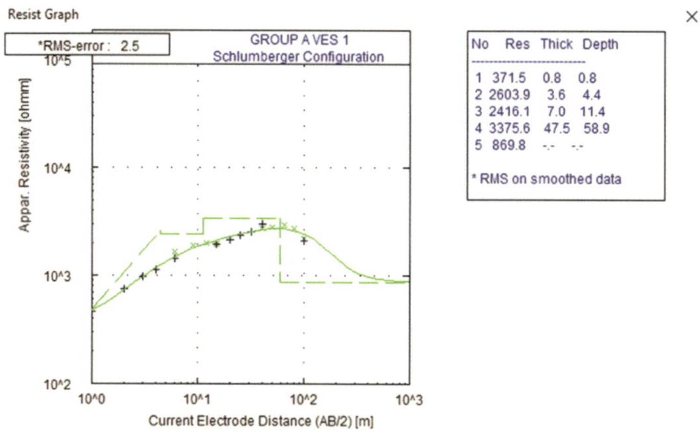

Figure A1.
Resistivity curve using Schlumberger configuration for vertical electrical sounding (VES) at site 1, showing apparent resistivity variations with depth.

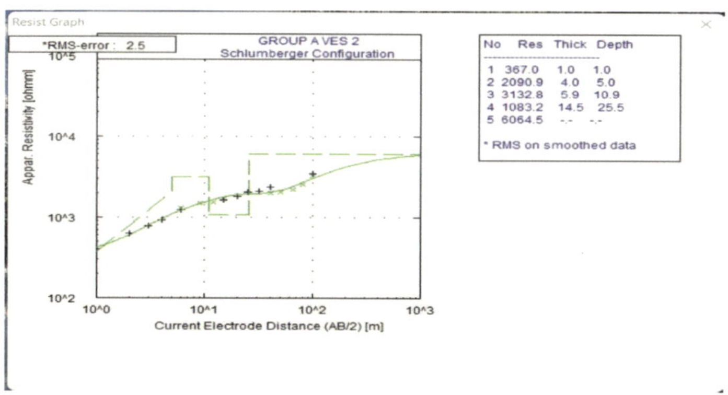

Figure A2.
Resistivity curve using Schlumberger configuration for vertical electrical sounding (VES) at site 2, showing apparent resistivity variations with depth.

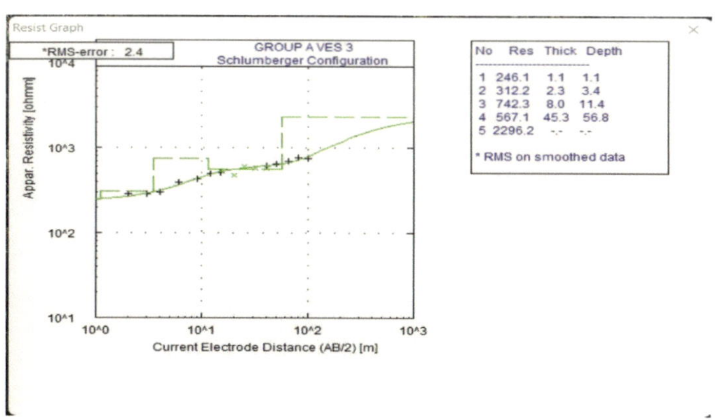

Figure A3.
Resistivity curve using Schlumberger configuration for vertical electrical sounding (VES) at site 3, showing apparent resistivity variations with depth.

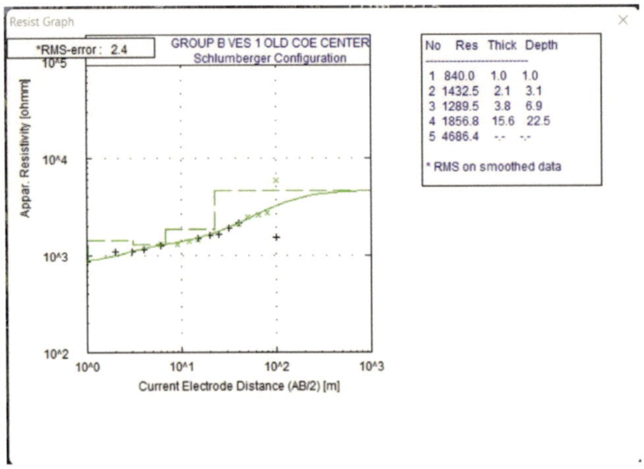

Figure A4.
Resistivity curve using Schlumberger configuration for vertical electrical sounding (VES) at site 4 (old COE center), showing apparent resistivity variations with depth.

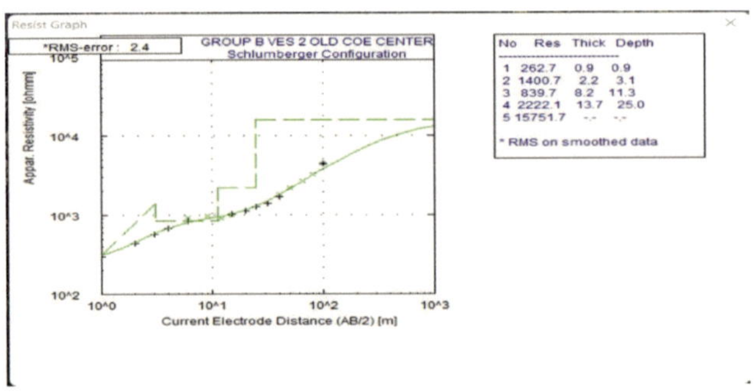

Figure A5.
Resistivity curve using Schlumberger configuration for vertical electrical sounding (VES) at site 5 (old COE center), showing apparent resistivity variations with depth.

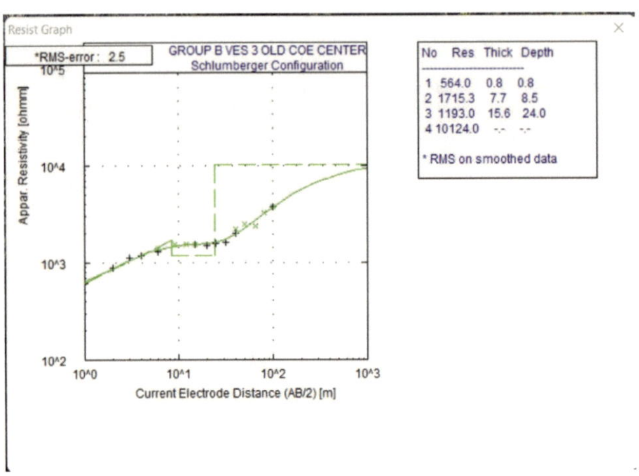

Figure A6.
Resistivity curve using Schlumberger configuration for vertical electrical sounding (VES) at site 6 (old COE center), showing apparent resistivity variations with depth.

Author details

Efetobore Gladys Maju-Oyovwikowhe[1*], Ekenakema Uwa-Igbinoba[2]
and Owens Monday Alile[3]

1 Faculty of Physical Sciences, Department of Geology, University of Benin, Benin City, Nigeria

2 Centre of Excellence in Geosciences and Petroleum Engineering, University of Benin, Nigeria

3 Faculty of Physical Sciences, Department of Physics, University of Benin, Benin City, Nigeria

*Address all correspondence to: efetobore.maju@uniben.edu

References

[1] Reynolds JM. An Introduction to Applied and Environmental Geophysics. 2nd ed. Chichester, UK: Wiley; 2011

[2] Loke MH, Chambers JE, Rucker DF, Kuras O, Wilkinson PB. Recent developments in the direct-current geoelectrical imaging method. Journal of Applied Geophysics. 2013;**95**:135-156

[3] Telford WM, Geldart LP, Sheriff RE. Applied Geophysics. 2nd ed. Cambridge, UK: Cambridge University Press; 1990

[4] Fisseha S, Demissie D, Nigussie W. Geophysical investigation for groundwater exploration in complex geological settings: A case study in Ethiopia. Journal of African Earth Sciences. 2021;**175**:104137

[5] Burger HR, Sheehan AF, Jones CH. Introduction to Applied Geophysics: Exploring the Shallow Subsurface. New York, USA: W.W. Norton and Company; 2006

[6] Garré S, Javaux M, Vereecken H. Advances in agrogeophysics: Monitoring soil moisture dynamics using multi-dimensional geophysical surveys. Geophysical Research Letters. 2023;**50**(4):e2023GL098912

[7] Egbo ID, Airen OI. Integrated geophysical techniques for subsurface characterization: A case study from Nigeria. Nigerian Journal of Geophysics. 2023;**27**(1):45-62

[8] Maju-Oyovwikowhe EG, Tahiru AK. Subsurface characterization using integrated ERT and seismic methods in a tropical environment. Journal of Environmental and Engineering Geophysics. 2021;**26**(3):231-243

[9] Ogungbe AS, Oluwaseun O, Adeyemi SA. Geophysical and geotechnical assessment of a proposed construction site in a sedimentary terrain, Nigeria. Engineering Geology. 2021;**291**:106241

[10] Kearey P, Brooks M, Hill I. An Introduction to Geophysical Exploration. 3rd ed. Oxford, UK: Wiley-Blackwell; 2002

[11] Alcalá FJ, Nieto JJ, Fernández I. Integrating ERT and seismic refraction tomography for sinkhole investigation in karst areas: A case study. Journal of Applied Geophysics. 2024;**212**:104513

[12] Uhlemann S, Chambers JE, Wilkinson PB. Subsurface characterization across hillslopes using integrated near-surface geophysical techniques. Hydrology and Earth System Sciences. 2023;**27**:311-326

[13] Olatunji O, Ajayi O, Akindele A. Geological Map of Nigeria. Abuja, Nigeria: Geological Survey Agency of Nigeria; 2014

[14] Dahlin T, Loke MH. Resolution of 2D Wenner resistivity imaging as assessed by numerical modelling. Journal of Applied Geophysics. 1998;**38**(4):237-249

[15] Alile OM, Ujuanbi O, Emenike EA. Geoelectric investigation of groundwater in Obaretin-Iyanomo area of Edo state, Nigeria. Journal of Geology and Mining Research. 2011;**3**(1):13-20

[16] Sharma PV. Environmental and Engineering Geophysics. Cambridge, UK: Cambridge University Press; 1997

[17] Short KC, Stauble AJ. Outline of geology of Niger Delta. American

Association of Petroleum Geologists Bulletin. 1967;**51**(5):761-779

[18] Jol HM. Ground Penetrating Radar Theory and Applications. 1st ed. Amsterdam, The Netherlands: Elsevier; 2009

[19] Vozoff K. The magnetotelluric method. In: Electromagnetic Methods in Applied Geophysics. Vol. 2. Tulsa, OK, USA: Society of Exploration Geophysicists; 1991. pp. 641-711

[20] Minsley BJ, Wellman TP, Wainwright HM. Geophysical monitoring of permafrost evolution using time-lapse electrical resistivity tomography in the Arctic. The Cryosphere. 2023;**17**:2045-2065

[21] Carrière SD, Doussan C, Chalikakis K. Impact of soil and subsoil conditions on tree response to drought: Insights from electrical resistivity tomography. Journal of Environmental Management. 2024;**331**:116006

[22] Emvoutou HC, Ndomè EEP, Diongue DM, Ndam JR, Stumpp C, Ketchemen-Tandia B, et al. Hydrochemical and isotopic studies providing a new functional model for the coastal aquifers in douala coastal sedimentary basin (DCSB)/Cameroon. Science of the Total Environment. 2024;**912**:169412

[23] Gnanachandrasamy G, Prabu G, Elangovan S, Thangarajan M. Remote sensing and GIS-based groundwater potential zone mapping in Ariyalur District, Tamil Nadu. Journal of the Geological Society of India. 2023;**92**(4):484-490

[24] Giménez-Forcada E. Space/time development of seawater intrusion: A case study using HFE-diagram in the eastern Spain coastal plain. Journal of Hydrology. 2024;**517**:61

[25] Fashae OA, Tijani MN, Ayanshola AM. Delineation of groundwater potential zones using integrated GIS and remote sensing techniques. Sustainable Water Resources Management. 2023;**9**:32-49

[26] Adams LH, Williamson ED. Elastic properties of minerals under high pressure. Journal of the Washington Academy of Sciences. 1951;**13**(5):413-428

[27] Salem HS. Hydraulic conductivity of shallow sediments from resistivity measurements using Schlumberger vertical electric soundings. Energy Sources, Part A: Recovery, Utilization, and Environmental Effects. 2001;**23**(7):599-618. DOI: 10.1080/00908310119202

[28] King IR. The structure of star clusters: III. Some simple dynamical models. Astronomical Journal. 1966;**71**:64

[29] Toksoz MN, Cheng CH, Timur A. Seismic studies and earth structures. Geophysical Research Letters. 1976;**3**(1):45-52. DOI: 10.1007/BF01878522

[30] Gardner GHF, Gardner LW, Gregory AR. Formation velocities and density: The diagnostic basics for stratigraphic traps. Geophysics. 1974;**39**(6):770-780. DOI: 10.1190/1.1440639

[31] Abd El-Rahman M. The potential of absorption coefficient and seismic quality factor in delineating less sound foundation materials in Jabal Shib Az Sahara area, northwest of Sanaa, Yemen Arab Republic. In: Egypt, M. E. R. C. Earth Science. Vol. 5. Ain Shams University; 1991. pp. 181-187

[32] Abd El-Rahman M. Evaluation of the kinetic elastic moduli of the surface

materials and application to engineering geologic maps at Maba-Risabah area (Dhamar province), Northern Yemen. Egypt Journal of Geology. 1989;**33**(1-2):229-250

[33] Birch F. Handbook of Physical Constants. Vol. 97. Geological Society of America Memoir; 1966. p. 613

[34] Gassman F. Seismische Prospektion. Stuttgart: Birkhaeuser; 1973. p. 417

[35] Tatham RH. Vp/Vs and lithology. Geophysics. 1982;**47**(3):336-344

[36] Sheriff RE, Geldart LP. Exploration Seismology. 2nd ed. Cambridge University Press; 1986. p. 316

Chapter 6

Overview of Groundwater Occurrence, Quality, and Challenges in Nigeria

Adamu Idris Tanko, Auwalu Yola Lawan and Rahama Darma Tijjani

Abstract

In Nigeria, the hydrostratigraphy of the country governs groundwater resources, which follow the occurrence and patterns of the aquifers, aquitards, or aquicludes. The country has two major hydrogeological units: the basement rocks and sedimentary basins. These units possess distinct characteristics in terms of groundwater potential. The basement complex consists of crystalline rocks that lie beneath the sedimentary layers, characterized by weathered, jointed, sheared, or faulted formations. The sedimentary terrains are majorly subaquatic or subaerial areas. Groundwater resources have been identified and discussed on majorly these two geological types and especially as they occur in each of the eight hydrological units demarcated in the country. The chapter also discusses various cases of groundwater quality across the country and concludes by highlighting the challenges associated with groundwater exploitation and management.

Keywords: groundwater, aquifer, recharge, basement complex, sedimentary formation, water quality

1. Introduction

In many parts of the world, groundwater resource is dictated by the hydrostratigraphy of the area. This explains the occurrence and patterns of aquifers, aquitards, or aquicludes. The basement complex rocks and the sedimentary formations are the two major hydrogeological units in Nigeria [1]. The basement complex is mainly made up of dioritic rocks, quartzites, granitic and migmatitic gneisses, and metasedimentary schists that have undergone mild metamorphism [2]. Overlying these rocks are sedimentary layers, which are cross-bedded, arkosic, gravelly, and poorly sorted, dating back to the Cretaceous and Tertiary periods. These sedimentary rocks are found in the southern, northeastern, and northwestern regions. The sedimentary layers are generally divided into three parts: a lower sequence of non-marine sandstones, siltstones, and mudstones; a middle section consisting of marine shales and limestones mixed with sandstones and siltstones; and an upper sequence of sandstones that were deposited in either continental or coastal environments [3]. These two geological types possess distinct characteristics in terms of groundwater potential and availability in every part of the country.

Groundwater is generally mostly explored more within the sedimentary formation where its quest is less difficult to achieve than the basement complex areas [4]. For instance towards the southern part of the country, especially in the Niger Delta and Anambra Basins that are underlain by sedimentary formations, the prospects of good aquifers increase in which groundwater yield becomes extensive to the extent that free flow conditions occur in confined areas [5]. This condition is different in "the high-plains of the Hausaland" in northern Nigeria [6].

The primary objective of the chapter is to provide empirical explanation that shall guide towards understanding the geography of groundwater occurrence for sustainable management in majorly the two geological types and in each of the eight identifiable river basins in the country. Also, the quality of the groundwater is explained. The challenges faced in groundwater management in the country are highlighted.

2. Material and methods

Material for the chapter is drawn from library search of empirical data about the groundwater situation in each of the basins that are underlain by both the basement complex rocks and sedimentary formations. The data must have been differently presented previously in disaggregated forms and authored by several researchers in the subject areas covering wide periods. Nigeria is demarcated into eight hydrological units. The characteristic occurrences in each of the units have been described and analysed. Similarly, data on the quality of groundwater in the country were also obtained through literature search and are presented under the different basins. Discussions are basically descriptive in nature. Finally, challenges of groundwater exploitation and management are highlighted based on experiences of the authors.

3. Nigeria's drainage pattern

Nigeria has two major rivers, the Niger and the Benue. River Niger, which has its origin from Guinea Highlands, is estimated to cover about 88,000 ha of floodplain while the River Benue flows from Cameroon Mountains and is estimated to cover 312,000 ha floodplain. The rivers are characterized by rapid and steeply slopes and overflow banks [7]. The two rivers meet at a confluence around Lokoja, at the centre of the country (**Figure 1**). From there, the rivers flow southwards to empty into the Atlantic Ocean through a network of creeks and distributaries, which form the Niger Delta. There are, however, several major tributaries within the larger Niger-Benue trough. These include the Sokoto-Rima in the northwest, the Kaduna in the central area, and the Gongola in the northeast. Others include the Anambra, Cross, Imo, and Kwa-Iboe, in the southeast/south-south and in the southwest are Ogun, Osun, Benin. In addition to these, Nigeria has a major inland water basin that drains into the Lake Chad at the extreme part of the northeast.

The geology of Nigeria is dominated mostly by the basin sedimentation (along the Niger Benue trough) that are mostly the older sedimentary rocks of the secondary age, while some aspect of the northern blocks which is the very old and highly altered basement complex rocks of the Precambrian [6]. Other minor formations are the volcanic rocks around the Jos Plateau and the river alluvium found along the major

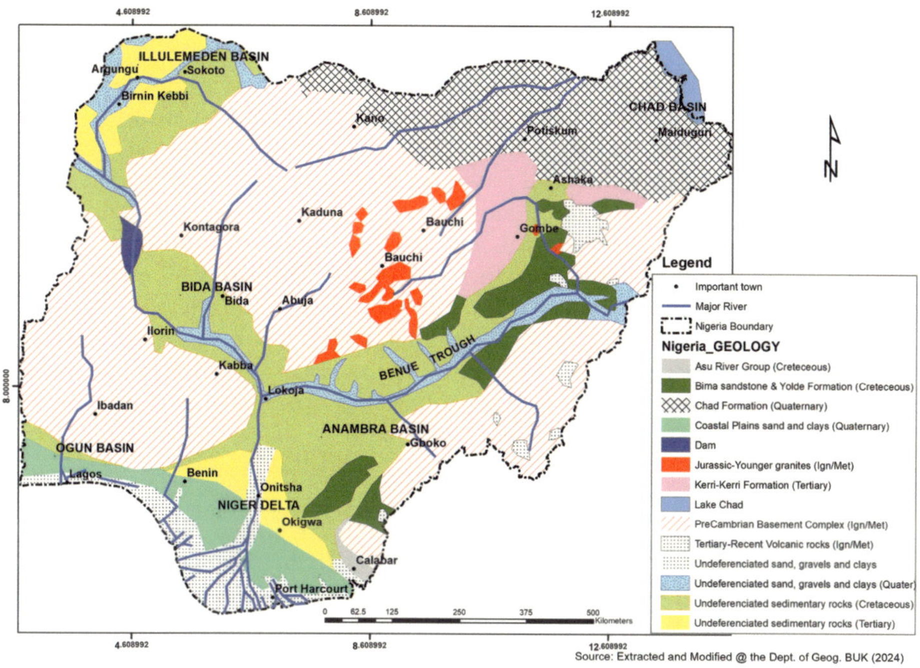

Figure 1.
Drainage basins and major towns in Nigeria.

rivers. These underlay the different river basins in the country which have been classified based on the following three hydro-meteorological regions [5]:

a. The Sahel Region: found in areas above latitude 11 degrees latitude and 3 to 8 degrees longitude. The area is characterized by low annual rainfall, ranging between 500 and 750 mm. It also experiences significant temperature extremes, with average temperatures reaching 44°C between April and May and dropping below 20°C in December and January. The region encompasses the Hadejia-Jama'are and Sokoto-Rima basins.

b. The Savanna Region: this zone experiences an average annual rainfall of 1000 to 1250 mm, with about 60 to 80 rainy days each year in the northern region. It encompasses five major river basins: the Kaduna River Basin, the Upper and Lower Benue River Basins, and the Upper and Lower Niger River Basins.

c. The Tropical Rainforest Region: this zone includes the Anambra, Cross-River, Kwa-Iboe, Niger Delta, and Southwestern Coastal basins. The main hydro-meteorological characteristics of these regions are the significant rainfall levels, which range from 1250 to 4000 mm, depending on the location, from the Anambra basin to the coastal basins.

4. Hydrogeology and groundwater occurrence in Nigeria

In Nigeria, aquifers are classified into two main types: basement fluvio-volcanic aquifers and sedimentary aquifers. According to Adewumi and Anifowose [8], groundwater availability in areas with basement rocks depends on either a thick soil overburden (known as overburden aquifers) or water-retaining fractures (referred to as fractured crystalline aquifers). Groundwater is stored in fractures and fissures within the weathered zones of igneous, metamorphic, and volcanic rocks, with thicknesses ranging from <10 - 60 m in both desert and humid rainforest regions [9]. In alluvial deposits, where Eolian and fluvial sediments are present, groundwater mainly fills pore spaces, unlike in sedimentary deposits where it may be found in different types of spaces. These alluvial deposits are considered primary aquifers because the water is stored in their original pore spaces. The variety of sedimentary layers in Nigeria's basins—differing in age, mineral content, and chemical composition—affects the amount and quality of groundwater [9]. The extensive sedimentary basins, which cover about 40% of the country, hold a large portion of the groundwater reserves. These basins contribute to around 67% of the estimated $51.93 \times 109\ m^3$ of potential annual groundwater resources [10].

4.1 Descriptions of the aquifer distribution in Nigeria

Here an attempt is made to describe the aquifer distribution in the eight river basins across Nigeria (**Figure 2**) relating each of them to the geological formation that underlay them. It is important to note that even as one may want to make a generalized scheme to suggest that the basins should be categorized under each geological type and the geographical sizes of each of the basins cannot allow such. However, discussions may heavily lay emphasis to the particular geology that predominates.

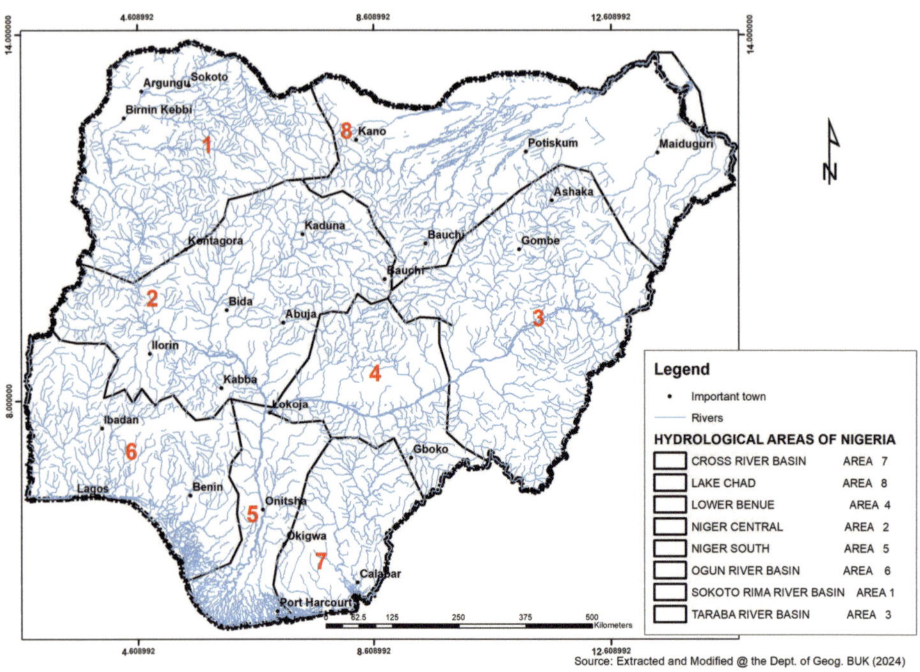

Figure 2.
River basins and their hydrological areas.

DOI: http://dx.doi.org/10.5772/intechopen.1006981

4.1.1 Sokoto-Rima basin

The Sokoto-Rima basin is mainly drained by the River Sokoto, with the Zamfara and Rima Rivers as its main tributaries. Groundwater comes from the fractured meta-sediments of the basement complex, which have shallow boreholes (30–40 m deep) and modest yields (0.5–0.8 lit/sec) [11]. At greater depths, where conditions might support artesian flow, the gravel, conglomerate, and grit layers of the Ilo/Gundumi Formation act as aquifers, though these are often limited by clay layers. Generally, the shallower aquifers are unconfined. In the Gusau-Sokoto Road area, yields of up to 8 lit/sec have been reported, with transmissivity reaching 978 m^2/day [12]. Typical yields usually range from 2 to 8 lit/sec, and transmissivity can vary from less than 10 to around 1000 m^2/day [13].

At Bakura Farm, boreholes average 85 m in depth and yield around 3 lit/sec [11]. Hydrogeologically, the Rima group, which includes the Taloka Formations with poor aquifer qualities and the Dukamaje, which acts as an aquiclude, is not particularly remarkable. The Wurno Formation in the Basin, on the other hand, is a moderate-output aquifer with a thickness of 5–30 m and a recharge area of 330 km^2 [14]. The Sokoto Group consists of the Dange (an aquitard), the Kalambaina (a perched aquifer that provides seasonal water), and the Gwandu Formation, which is the most prominent aquifer in the Sokoto Basin. The Gwandu Formation features two main aquifers: a lower sandy aquifer with limited capacity and an upper sandy aquifer. In the water table aquifer, average water levels are around 21 meters deep, though they can reach up to 100 meters in the uplands near the border with the Niger Republic. The thickness of this aquifer ranges from 13 to 60 m and is further confined to the west by 20 m of lignite and clay, which creates artesian conditions [15]. In the unconfined areas of the Sokoto Basin, the water table typically ranges from 15 to 75 m deep [16]. Towards the eastern edge of the Basin, especially around Argungu, artesian conditions are found in limited aquifers at depths of 75 - 100 m. However, piezometric levels decrease to about 50 m below the surface as you move further west [14, 17].

4.1.2 The Niger central

This comprises both the Upper and Lower Niger Basins. In the upper portion, the Basin is charged with water from mainly the Sokoto-Rima and also Kaduna River systems. Most of the tributaries of these rivers flow over long distances from their source areas, usually located in the more humid areas of the Jos–Katsina–Zaria north central massif, typified by the Jos Plateau. The Kaduna River is a major tributary of the Upper Niger, originating in the Kajuma Hills in Nigeria's Jos Plateau. It travels about 200 km to Kaduna town and then continues for another 100 km into the areas where the Shiroro Dam is, eventually flowing into the Niger River near the northern shores of Pategi [18]. The river flows through Kaduna town on both the north and south sides [19]. Over 80% of the upper section of the basin is covered by basement complex rock, which has shallow depths and yields between 0.2 and 1 lit/sec [5].

In this northwestern sector of the basin, only the Illo Formation is exposed, though detailed mapping of the area has not been carried out [13]. Oteze [20] asserts that the aquifers only appear as lenses within primarily clayey strata. Water yields are hence often low, despite the presence of restricted conditions in some strata. The fine texture of the sands and the quick alternation of sands with clay mudstones and silt in the eastern sedimentary region of the basin contribute to lessen the Nupe Sandstone's groundwater potential. However, according to Preeze and Barber [21], there is a lot of groundwater present in the Nupe sandstone regions, with a water table that varies from 0 to 100 meters. The eastern sedimentary basin extends into the Lower Niger basin and possibly merges with the Lower Benue Basin around Rivers Benue – Niger confluence areas [13].

The lower portion of the basin is basically alluvial deposits associated with the course of the River Niger. The alluvial deposits are variable in texture, composition, and thickness, ranging from coarse sands, gravels to fine silts, and clays. Invariably, they overlie the Nupe Sandstone or its equivalent, and in places they cover the basement directly. Groundwater in the alluvial deposits is mainly recharged by rainfall and floodwaters from nearby rivers [13]. These alluvial layers help sustain substantial groundwater flow underground. In the Koton Karfe area, alluvial deposits have been used for boreholes, although detailed records are lacking [13]. In the southeastern regions, including Kontagora, Mokwa, and Bida, the Nupe Sandstone is the predominant layer [22]. This sandstone formation includes thin layers of carbonaceous shale and clay mixed with sandstones and siltstones, which are somewhat cemented. The Nupe Sandstone typically yields between 1.8 and 4 lit/sec [21].

4.1.3 Gongola basin

Located in the higher regions of the Benue Valley, this area is locally referred to as the Upper Benue Basin. The River Benue is fed by numerous streams and rivers from both the north and south, forming its main drainage system. Key river systems include the Faro and Taraba from the south and the Gongola, Kilunga, and Pai from the north [13]. The upper part of the basin features patches of basement complex rocks, including volcanic plugs and basaltic flows, as well as some Cretaceous sedimentary rocks. The geological base of the area is comprised of a thick layer of Asu River group shales, continental sandstones from the Bima sandstone, and additional shales, clays, and limestones [13]. The Zambuk Ridge separates the Upper Benue from the Chad Basin. The basement complex rock in this area is crystalline and has limited water-yielding capacity. However, weathered basalts in the Sugu Plateau, Longuda, and Biu regions are of some hydrogeological interest. More significant is the Bima sandstone found in the Yola, Jimeta, and Dabore areas, which is feldspathic and contains grit, pebbles, and clay beds. Its hydrological performance is constrained by its highly cemented nature and rock-like characteristics. Reports indicate that it yields around 2–8 lit/sec [13]. The Yolde Formation overlies the Bima sandstone and consists of about 152 m of thinly bedded sandstones, followed by alternating mudstones and shelly limestone. The Yolde Formation underlies the two sedimentary sub-basins of the Upper Benue. It is a more promising water bearing formation than the Bima sandstones as the sandstones and limestone are more permeable. Yields of up to 4 lit/sec have been obtained at Gombe and Numan [13]. The water normally occurs under water table conditions.

4.1.4 Lower Benue river basin

This basin covers the lower part of the Benue Valley, stretching from the River Wase in the Jos Plateau to where the Benue and Niger rivers meet. It is drained by several river systems, including the northeast-flowing River Wase, Shemankar, Dep, and Mada, as well as the Donga, Bantaji, and Katsina Ala, which flow in the opposite direction. The River Benue serves as the main outlet for these river systems.

The Lower Benue and Upper Benue regions have similar geological formations due to their comparable lithology [13]. These formations include the basal sandstones of the Lafia, the sandstones of the Makurdi and Ezeakku Formations, the Awgu Formation, and the interbedded sandstones of the Awe Formation. Several boreholes have been drilled into the Makurdi Formation with varying degrees of success, yielding between 2.5 and 8.7 lit/sec [13]. In Assakio, north of Lafia, a 150 m deep borehole drilled into the Awgu Formation achieved artesian flow. The junction where the Lafia Formation meets the underlying Awgu Formation produces multiple springs due to the highly permeable sandstone in this area, which ranges from 10 to 15 m in thickness [13].

4.1.5 The Niger south basin

This comprises the two basins that are locally known as the Anambra and the Niger Delta. The Anambra Basin is one of Nigeria's most important hydrogeological basins, taking after the Sokoto and Chad Basins. It is drained mainly by the Anambra River and its main tributaries are the Mamu and Adada. The most important aquiferous formation in the Anambra Basin is the Ajali Formation. This formation is underlain by 585 m of the Mamu Formation and Nkporo Shale [13]. The Ajali Formation, known for its substantial water-bearing properties, has indeed been a source of prolific artesian flows in areas like Umumbo, Mbaguowa, and Aguobuowa, located on the outskirts of the basin. According to Offodile [13], in the Awka region, the aquifer associated with the Ajali Formation could potentially be encountered at much greater depths, ranging from about 360–800 m. This variation in depth highlights the geological complexity and the potential for significant groundwater resources within this formation. Apart from the Ajali Formation, shallow aquifers identified in Awka area of the Mamu Basin include the Ugwuoba sandstone also described as Ebenebe sandstone and the Amenyi sands known to outcrop at the eastern outskirts of Awka town. All these aquiferous sand bodies are members of the Imo Shale [13]. In the Ameki Formation, the sandstones are generally very permeable. In Onitsha area, the water table is high about 20–30 m [13]. The same source submits that in Nanka sands around Nanka, Idemili, and Oko, the water table is generally very low ranging from 30 to 300 m in depth.

The Niger Delta, on the other hand, is primarily a sedimentary basin where groundwater recharge is mostly through infiltration and from precipitation and rivers/lagoons at the surface. The Benin Formation and the Deltaic Formation are the primary aquifers in the Niger Delta, playing a crucial role in water supply for the region. Most boreholes in the northern parts of the Basin tap into these extensive and largely unrestricted aquifers, which provide reliable sources of groundwater. In certain locations, unconfined aquifers with artesian groundwater flow have also been identified, indicating that the water is under pressure and can rise above the level of the aquifer without pumping. Geological logging from existing boreholes has revealed that these formations contain deep sandy units at specific depths in some areas, which are crucial for understanding the distribution and accessibility of groundwater resources. These sandy units likely contribute to the high permeability and water-bearing capacity of the aquifers, making them significant sources for water extraction in the Niger Delta [23]. The Deltaic Formation is classified as an unconfined aquifer, which is the first and topmost groundwater unit recharged directly by infiltration from precipitation and base flow. The water table in the Basin is only 0 to 9 m below the surface of the ground [24]. Although part of the rainfall is lost through runoff and evapotranspiration, this area's significant rainfall, which ranges from about 2400 mm inland to 4800 mm along the coast, results in little variation in the water table [24]. The Deltaic Plains have specific capacities ranging from 160 to 320 m per d/m, according to Offodile [13]. The sediment composition of the Benin Formation makes it more permeable than the Deltaic Plains, with a specific capacity ranging from 150 to 1400 m per d/m. This region often has a water table that is 3–15 m below the surface. The water level can drop by 2.1–3.6 m due to seasonal variations [24]. Confined aquifers are found in both the Deltaic and Benin Formations, and they usually yield somewhat high artesian flows. In certain areas, these aquifers are covered by a 36 m thick clay layer, with the aquifers lying approximately 100 m beneath them. Hydrogeological studies indicate a link between the confined aquifers near the coast and the unconfined aquifers of the Benin Formation farther north. As you approach inland, the aquifers thicken, while the clay layers that surround them thin. The specific capacity of these formations ranges between 90 and 320 m per d/m [24]. Restricted aquifers are found in the southeastern part of the Basin, where the Benin Formation is located. These aquifers are confined by several clay layers and are mainly made up of very coarse- to medium-grained sands. The specific capacity of these aquifers ranges from 140 to 180 m per d/m [24].

4.1.6 *Ogun river basin*

The Ogun Basin is a major coastal basin in southwestern Nigeria, drained by the Ogun River, which originates in the Iran Hills at about 530 m above sea level and flows south for 480 km into the Lagos Lagoon [25–27]. The main tributaries are the Ofiki and Opeki Rivers. Although the Ogun Basin is important, comprehensive hydrogeological data is scarce, with limited information on individual boreholes. Most boreholes in the Basin reach into the Pre-Cambrian Basement Complex. In the Ogun Basin, the yields from boreholes are not extensively recorded. Near the base, however, an artesian flow of 90–135 lit/hr was supplied by GSN. BH. No. 2614 at Ewekoro [13]. Another borehole, GSN. BH. No. 1583 at Itori, had an artesian flow at 81 m below the surface, with a specific capacity of 92.7–45,000 lit/hr/m and an anticipated output of 450 lit/hr [13]. More thorough hydrogeological investigations are required for a thorough assessment, even though some boreholes exhibit significant water production. One of Nigeria's eight hydrological areas, the Western Littoral Hydrological Area HA-6, contains the Osse-Owena Basin, commonly referred to as the Benin-Owena Basin. This Basin has not been fully explored hydrogeologically.

4.1.7 *Cross-river basin*

The Basin is locally divided into the Kwa-Iboe and Cross-River Basins. The Kwa-Iboe Basin features three main hydrostratigraphic units: the Alluvial Deposit, Coastal Plain Sand, and Lower Sand Aquifers [28, 29]. The Alluvial Deposit, the largest unit, is separated from the Lower Sand Aquifer by the Imo Shale and varies in thickness from 35 m in the north to 200 m in the south. It is mostly semi-confined or unconfined, particularly in the southern part [30]. Groundwater occurs principally under water table conditions. However, data from some settlement including Akata, Orlu-Ezima, Umuahia, Atta, and Awomama Umuma circle, where the depth of the water table fluctuates from about 50 m on the circle's periphery to over 75 m towards the centre of the Orlu Ridge, indicate that the water table is very low in some regions. The depth of the water table in the region south of Owerri and Aba increases at the lower plains region (from 6 m along the shore to 30 m inland) [13]. The predominant groundwater flow into the Atlantic Ocean is in a north-to-south direction. In the Imo River and the Cross River, there are variances in the northeast-southwest and northwest-east directions as well [30].

The Cross-River Basin, on the other hand, is described as a hydrogeologically problematic groundwater basin, which according to Offodile [13], the Asu River Group of the Cross River is intensely fractured and large quantities of saliferous groundwater were pumped out during lead - zinc mining activities (GSN RPT. No. 1429). In less fractured situations, the Asu River Group cannot be considered for groundwater exploitation. However, where fractured, as often the case in the Abakaliki districts, groundwater occurs in reasonable quantities but with steep drawdowns because of the limitations in the extent of the reservoirs caused by impermeable shale-clay boundaries. Only the coarser, less-cemented sandstones and fractured limestones with solution channels and voids contain the aquifers of the Ezeaku Formation. In contrast, the Awgu Formation's sandstones are thin and typically of restricted extent, which results in low yields [13]. Even worse as groundwater reserves are the Nkporo and Mamu Formations that consecutively overlie the Awgu Formation. All of these extremely weak aquiferous formations, which underlie the majority of the Cross-River Basin, have contributed to the basin's overall hydrogeological issues. The Ajali and Benin Formation outcrops in the districts of Utapate, Etinan, Oron, and Calabar are among the more useful younger aquifers that are present beneath the extreme southeast corner, where the Cross-River drains into the sea [13].

4.1.8 The lake Chad basin

The Basin is the largest inland drainage basin in Africa, spanning parts of Nigeria, Chad, the Central African Republic, and Cameroon. In Nigeria, its main tributaries are the Hadejia, Jama'are, Komadugu, and Yobe Rivers, which originate from the Jos Plateau and flow through Kano, Yobe, Bauchi, and Borno, moving from high rainfall areas towards the drier Lake Chad region. The Basin's geology includes a mix of rock formations: the Basement Complex, Younger Granite Complex, Gundumi Formation, Kerri-Kerri Formation, and Chad Formation. In the Basement Complex and Younger Granite areas, boreholes are typically shallow (20-35 m) with specific capacities of about 8–9 lit/m/min [13]. In the Gundumi Formation, groundwater is found under water table conditions, with coarse sands providing good aquifers, except in areas where the sediments are too thin to hold significant amounts of water, particularly near the Basement inliers. Boreholes in the area average 45 meters deep, with water yields ranging from 3.2 to 6.25 lit/sec [5].

The Kerri-Kerri Formation, up to 200 m thick, is poorly explored and has deep water levels, while the Chad Basin sediments range from 115 to 132 m thick with yields of 3.3–5 lit/sec [13]. The aquifer layers in the Chad Formation are not clearly defined in this region. The Hadejia-Yobe Basin's upper 20–30 m of alluvial sands store significant water, recharged directly from river flows [5].

Most of the rivers flowing into Lake Chad originate from the watershed areas of the Jos Plateau and the Adamawa highlands. Major rivers include the Hadejia-Yobe, Alo, and Yedsaram from the southwest, and the Bambassa, Chari, Illi, and Lagone from the southeast. The Yedseram-Ngadda River flows from central Nigeria to the northwest, along with its minor tributaries, eventually emptying into Lake Chad. The Hadejia-Yobe, Yedseram, and other tributaries are seasonal, with the Yedseram starting to flow slightly later than the Hadejia-Yobe after the rainy season begins. The arid northern regions contribute little to no surface flow [13]. Hydraulically, the

Kerri-Kerri and Chad Formations are significant in the Chad Basin. The Kerri-Kerri Formation is too deep to be of widespread hydrogeological interest, while the Chad Formation is well studied and consists of three main aquifers:

a. An upper aquifer located at a depth of 30–100 m

b. A middle aquifer, found primarily in the eastern part of the basin, with a thickness of 40–100 m, occurring from around 230 m deep near Maiduguri.

c. A lower aquifer made up of 100 m of medium to coarse sands and clays, situated at a depth of 425–530 m.

In Maiduguri, the upper aquifer extends up to 105 m deep, composed of sands and clays, with water levels between 10 and 15 m and yields of 2.5–30 lit/sec [5]. The middle aquifer, separated from the upper by 150 m of clay, is 300 m thick and highly confined beneath additional 120 m of clay and shale layers [5]. This aquifer is transboundary, with recharge areas outside Nigeria. Due to overexploitation of the aquifer in the Chad Basin, groundwater levers have recently declined, leading to the need for deeper drilling to access the lower aquifer [31]. Isotopic evidence shows that the middle and lower aquifers contain ancient water, around 20,000 years old, not replenished by modern sources [32, 33].

5. Groundwater quality issues

In Nigeria, there have been numerous investigations on the quality of groundwater, albeit they have often been conducted locally and with few chemical elements. Around the country generally, groundwater quality is impacted by common contaminants (chemical, physical, and biological parameters), anthropogenic factors (urbanization, mining, industrialization, and geology), and natural impacts (hydrogeology and geology). Since groundwater is a byproduct of geological formations, various researches looked at the impact of geology on its quality. Where the aquifers are shallow, the resource is susceptible to contamination from household, agricultural, and industrial waste [34]. In other places, groundwater is affected by several factors including oil spills. A notable example is provided by UNEP in 2011 in their study on the environmental evaluation of Ogoni land [35]. Natural water quality issues have also been reported, particularly in the context of the Benue Basin's mineral exploration [36–38]. In the coastal regions of Nigeria, problems with saltwater intrusion have been caused by overexploitation of coastal aquifers, tidal impacts, and shallow wells. In this way, it can be summarized that groundwater quality is determined by both natural and human activities [39]. The hydrological conditions and geochemical and physical processes all determine the nature, amount, and duration of human activity on groundwater quality [40].

5.1 Pollution by heavy metals

Quality issues are known to predominate differently in different basins in the country. Across the Nigeria's industrial basins groundwater pollution by heavy metals, including iron (Fe), lead (Pb), cadmium (Cd), arsenic (As), mercury (Hg), chromium (Cr), and zinc (Zn), can directly contaminate groundwater through leaching from contaminated soils, industrial discharge, mining activities, and improper waste disposal. Heavy metals are the major problem due to the considerable mining activity, industrial discharge, and poor waste management procedures. They can cause serious health effect with different symptoms depending on the nature and quality of the metal ingested [41]. Research has revealed increased concentrations of heavy metals such as lead, cadmium, and arsenic in groundwater in a number of different areas, such as the Jos Plateau in the central part of the country and the Niger Delta in the south-south [42].

Mining activities in regions such as Zamfara (in the Sokoto Basin) and the Jos Plateau have led to significant heavy metal contamination of groundwater. Studies in the Zamfara artisanal gold mining have resulted in severe lead poisoning, affecting both water quality and public health [43]. According to other reports [44, 45], a variety of waste items in metropolitan areas are sources of heavy metal contamination. Water supplies and the environmental quality will deteriorate quicker in areas where heavy metal concentrations are high. Health concerns and hazards are unavoidable once the environment and water resources are contaminated with excess heavy metals. Industrial activities around areas like the Ajaokuta Steel Company, near the Niger-Benue Confluence have contributed to heavy metal contamination in groundwater, with high levels of chromium, nickel, and lead detected in local water sources [46].

5.2 Pollution by hydrocarbons

Other pollutants in Nigeria's groundwater are in the forms of hydrocarbons especially the polycyclic aromatic hydrocarbons (PAHs), iron (Fe), fluorides (F-), nitrates (NO_3), and sulphates (SO_4^{2-}). The presence of all these in groundwater primarily is due to mining, industrial activities, urbanization, and especially waste disposal systems. Nigeria, a leading global crude oil producer, transports petroleum products via pipelines to various depots across the country, from where they are distributed by tankers to consumers [47, 48]. The Niger Delta region, however, has experienced frequent oil spills caused by pipeline leaks, sabotage, and operational failures. These spills have resulted in significant hydrocarbon contamination of surface water and groundwater [49–51]. The Bodo oil spill in 2008–2009 was a notable incident, where large volumes of crude oil were released into the environment, severely impacting water quality and local communities [35]. Refineries and petrochemical industries discharge effluents containing hydrocarbons into water bodies. Similarly, inadequate waste management practices contribute to groundwater contamination. For example, the Warri Refinery and Petrochemical Company released effluents into the surrounding environment, affecting water quality [46]. Illegal artisanal refining activities, prevalent in the Niger Delta, involve crude methods of refining crude oil, leading to spills and the release of hydrocarbons into the environment. These activities often occur near water bodies, resulting in direct contamination of groundwater sources [52].

5.3 Other pollutants

Other pollutants of groundwater sources include fluorides. Fluorite is liberated during the weathering process and is absorbed into the groundwater system by infiltration. **Table 1** indicates that 47% of the water in the basement aquifer has fluoride levels below the acceptable limit of 1.00 mg/L, while 53% exceed this level [54]. Fluoride concentrations of 10.30, 7.72, and 8.0 mg/L were measured in a hand-dug well at Langtang and springs at Furzi and Dorong in the north-central

Basin	Formation (Aquifer)	T(°C)	pH	EC (μS/cm)	TDS (mg/l)	TH (mg/l)	Na^+ (mg/l)	K^+ (mg/l)	Ca^{2+} (mg/l)	Mg^{2+} (mg/l)	Cl^- (mg/l)	HCO_{3-}	SO_4^{2-}
Niger Delta	Benin	28.29	5.76	149.76	77.9	24.60	10.38	3.36	5.48	2.60	18.83	28.20	6.37
	Ogwashi Asaba	27.74	5.91	258.31	85.04	35.92	9.03	1.33	7.30	4.23	30.65	30.49	9.69
Benue Trough	Nanka Sands		5.50	313.25	153.75	163.33	0.00	3.05	39.50	15.50	53.25	13.50	8.00
	Enugu Shale		5.31	635.36	413.92	98.91	80.26	23.16	11.56	17.19	82.55	119.10	79.37
	Agla		6.61	625.38	372.53	0.01	0.25	0.02	46.40	19.29	23.42	237.79	35.75
	Asu River	28.76	6.49	539.18	283.78	316.70	97.20	20.80	50.00	46.01	59.56	192.08	87.38
Sokoto-Rima	Gwandu		5.89	180.82	117.84	67.06	57.84	9.31	16.99	3.97	13.71	27.09	7.50
	Kalambaina		7.05	364.71	402.50	106.80	202.09	52.29	55.60	14.89	17.27	156.93	25.52
Gongola & Lower Benue	North Central	26.46	6.13	185.87	136.32	80.95	12.27	4.07	18.57	8.29	12.00	81.38	42.69
	Adamawa Hills	30.79	6.64	334.07	320.03	119.56	27.68	9.25	30.59	10.34	8.06	183.06	5.09
	South West	28.07	6.57	377.36	177.72	88.72	27.88	8.46	24.30	6.71	22.59	92.98	14.96
	Oban Massif	28.62	5.91	123.68	93.39	27.44	2.63	3.62	6.23	2.84	1.89	18.43	6.09
	Obudu Plateau	27.44	5.89	199.03	127.59	107.61	3.98	2.67	37.41	3.38	19.86	47.27	58.62
Niger South Basin	Benin	27.64	6.68	4551.33	2395.44	790.16	508.03	46.58	134.09	109.18	1206.68	187.86	114.79
	Makurdi sandstone	31.85	6.83	27722.0	17641.25	1271.44	6136.75	243.15	390.58	70.80	10278.75	406.05	
	Awe	35.70	7.06	15441.2	9119.40	660.70	3371.00	110.44	201.58	37.62	4814.60	524.10	17.17
	Asu River	29.12	6.58	50203.91	33715.14	3875.83	11528.27	283.03	1216.95	200.03	20147.18	219.85	179.36
WHO (1993)	STANDARD		6.5–8.5	1500	1000	500	200	30	200	150	250	380	400

(Source: Ref. [53]).

Table 1.
Overview of physicochemical characteristics of groundwater for some basins in Nigeria.

basement complex [55, 56]. In the northeast basement aquifers, fluoride levels of 7.00 mg/L were found in both a borehole and a well at Dass [57]. Biliri, Kaltungo, and Shongwom are other areas with high values of fluoride concentration [54]. Values of 2.22 and 2.52 mg/L of fluoride concentration in the northwest basement aquifers were also recorded [53].

6. Challenges of groundwater management in Nigeria

Groundwater occurrences and management in Nigeria face several challenges, including:

1. Over-Extraction: Excessive pumping of groundwater for agriculture, industry, and domestic use can lead to the depletion of aquifers, causing water scarcity and land subsidence.

2. Climate Change: Altered precipitation patterns and increased evaporation due to climate change can affect the recharge of aquifers, potentially reducing groundwater availability.

3. There is a limited understanding of the primary aquifer systems and their potential for development.

4. The geological complexity results in development challenges such as failed wells and low aquifer yields, as well as a poor understanding of groundwater recharge and discharge processes.

5. There is limited information on how much groundwater is used by different sectors.

6. Water quality issues in certain regions arise from natural processes, as well as contamination from sewage, industrial runoff, and agricultural chemicals.

7. Decreasing of borehole yields and water levels result from overuse, along with high costs associated with groundwater development and the ongoing expenses of running and maintaining boreholes.

8. Inadequate or fragmented data and information systems hinder effective planning and monitoring.

9. Poor construction of hand-dug wells and lack of infrastructures and maintenance of water abstraction points.

10. Corruption associated with issuing drilling contracts, limited financial resources, inadequate monitoring networks, poor institutional arrangements, and policies to manage groundwater resources effectively.

11. Urbanization: Rapid urban growth often leads to increased demand for groundwater, putting pressure on already stressed aquifers.

12. Legal and Regulatory Issues: Weak enforcement of groundwater regulations and property rights can result in unregulated drilling and overuse. Addressing these challenges requires sustainable groundwater management, improved infrastructure, and increased awareness of the importance of protecting this vital resource.

7. Conclusion

Groundwater is essential for providing reliable and economical water supplies in both urban and rural areas. It includes water found in the unsaturated zone (soil), above the water table (capillary fringe), and below the water table. The availability of groundwater has become increasingly important due to the limited and declining yields of surface water throughout the year. Nigeria possesses significant groundwater reserves in both the basement complex and sedimentary aquifers, with sedimentary basins generally offering better aquifer conditions. In contrast, groundwater storage in the crystalline basement complex is relatively limited, and the failure rate for new boreholes in this region can reach up to 80% [53].

The practice of hydrogeology in Nigeria is hindered by a lack of appropriate hydrogeological maps, limited geological knowledge of the terrain, insufficient infrastructure, and the absence of effective legislation. These issues complicate the exploration, exploitation, operation, control, and management of Nigeria's groundwater resources. Sustainable development of these resources relies on a thorough understanding of groundwater availability and the processes of recharge and renewal.

Author details

Adamu Idris Tanko[1*], Auwalu Yola Lawan[2] and Rahama Darma Tijjani[2]

1 Department of Geography, Bayero University Kano, Nigeria

2 Department of Geology, Bayero University Kano, Nigeria

*Address all correspondence to: aitanko.geog@buk.edu.ng

References

[1] Ayobami IO, Theophilus AA. Groundwater occurrence and flow in varying geological formations. IOP Conference Series: Earth and Environmental Science. 2022;**1197**:012009

[2] Rahaman MA. Review of the basement geology of southwestern Nigeria. In: Kogbe CA, editor. Geology of Nigeria. Nigeria: Rock View Limited; 1989. pp. 39-56

[3] Petters SW. Central West African Cretaceous – Tertiary Benthic Foraminifera and Stratigraphy. Vol. 179. Stuttgart, Germany: Palaeontographica Abt A; 1982. pp. 1-104

[4] George NJ, Umoh JA, Ekanem AM, Agbasi OE, Jamal A, Thomas JE. Geophysical-laboratory data integration for estimation of groundwater volumetric reserves in coastal hinterland through optimized interpolation of interconnected geo-pore architecture. Journal of Coastal Conservation. 2022;**26**:56

[5] Joseph IA. Threats to water resources development in Nigeria. Journal of Geology and Geophysics. 2015;**4**:205

[6] Udo RK. Geographical Regions of Nigeria. California: University of California Press, Berkeley & Lost Angeles; 1970

[7] Tanko AI, Badamasi MM, Momale SB, Yusuf HE, Danjuma MN, Sanusi MM, et al. Access rights over land and water resources and implications on sustainable agriculture. In: The Niger-Benue Basins, a Report of Field-Based Research Funded under the National Research Fund (NRF) (2019-2021). Nigeria: Tertiary Education Trust Fund (TETFund); 2021

[8] Adewumi AJ, Anifowose YB. Hydrogeologic characterization of owo and its environs using remote sensing and GIS. Applied Water Science. 2017;**7**:2987-3000. DOI: 10.1007/s13201-017-0611-8

[9] Eduvie MO. Borehole failures in Nigeria. In: Paper Presented at a National Seminar Held on the Occasion of Water Africa Exhibition (Nigeria 2006) at Eko Hotels & Suites, Victoria Island, Lagos, on 15th November, 2006. Nigeria. 2006

[10] FMWRRD. The Study on the National Water Resources Master Plan. Federal Republic of Nigeria, Abuja: Japan International Cooperation (JICA) and Federal Ministry of Water Resources and Rural Development; 1995

[11] Wali SU, Umar KJ, Abubakar SD. Hydrochemical characterization of shallow and deep groundwater in basement complex areas of southern Kebbi state, Sokoto Basin, Nigeria. Applied Water Science. 2019;**9**:169. DOI: 10.1007/s13201-019-1042-5

[12] Anderson HR, Ogilbee W. Aquifers in the Sokoto basin north Western Nigeria, with a description of the general hydrogeology of the region. Water Supply Paper. 1973:1757-1L79. DOI: 10.3133/wsp1757L

[13] Offodile ME. Groundwater Study and Development in Nigeria. Jos, Nigeria: Mecon Services Ltd; 2002. pp. 30-36, 76-103

[14] Adelana SMA, Olasehinde PI, Vrbka P. A quantitative estimation of groundwater recharge in parts of Sokoto basin, Nigeria. Journal Environmental Hydrology. 2006;**14**(5):1-17

[15] Nwankwoala H. Hydrogeology and groundwater resources of Nigeria. New York Science Journal. 2015;**8**:89-100

[16] Adelana SMA, Vrbka P. Hydrogeological and isotopic research in the semi-arid area of northwestern Nigeria. In: Proceedings of the Biennial Groundwater Conference. Pretoria, Gauteng, South: Division of the Geological Society of Southern Africa CSIR Conference Centre; 7-9 Mar 2005

[17] Adelana SMA, Olasehinde PI, Vrbka P. Isotopes and geochemical characterization of surface and subsurface waters in the semi-arid Sokoto basin, Nigeria. African Journal of Science and Technology. 2003;**4**(2):76-85

[18] Emere MC, Dibal DM. Metal accumulation in some tissues/organs of a fresh water fish (*Clarias gariepenus*) from some polluted zones of river Kaduna. Journal of Biology, Agriculture and Healthcare. 2013;**3**(1):2013

[19] Al-Amin MA. Energy production and environmental concerns in Nigeria: The case of Kaduna petroleum refinery on its host communities. Journal of Energy Technologies and Policies. 2013;**3**:69-77

[20] Oteze GE. The hydrogeology of North Western Nigerian Basin. In: Kogbe CA, editor. Geology of Nigeria. Ibadan, Nigeria: Elizabeth Pub. Co; 1976

[21] Preeze D, Barber W. The distribution and chemical quality of groundwater in northern. Geological Survey of Nigeria Bulletin. 1965;**36**:1-20

[22] Adelana SMA, Macdonald MAS. Groundwater research issues in Africa. In: Adelana S, Macdonald M, editors. Applied Groundwater Studies in Africa, IAH Selected Papers on Hydrogeology. Vol. 13. London, New York, Leiden: CRC Press, Taylor and Fracis Group; 2008

[23] Obomighie PO, Gwegwe JO, Oluwadebi AG. Petrographic studies and paleoenvironmental reconstruction of some outcropping sediments in parts of eastern Dahomey basin, southwestern Nigeria. IOSR Journal of Applied Geology and Geophysics. 2016;**4**:9-15

[24] Ibrahim H, Robert MK, Christopher JW, Jamiu AA. Hydrostratigraphy and hydraulic characterisation of shallow coastal aquifers, Niger delta basin: A strategy for groundwater resource management. Geosciences. 2019;**9**(11):470

[25] Adeyemi O, Olutoyin F, Olumide O. Downstream hydraulic geometry across headwater channels in upper Ogun River basin, southwestern Nigeria. Africa Geographical Review. 2020;**39**:345-360

[26] Eludoyin AO, Adewole AO. A remote sensing based evaluation of ungauged drainage basin in southwestern Nigeria. International Journal of River Basin Management. 2019;**18**:307-319

[27] Olusola A. Morphologic and hydraulic variability of small bedrock and alluvial channels in relation to lithologic controls, upper Ogun river basin, southwestern Nigeria. Physical Geography. 2020;**41**:537-557

[28] Edet AE. Groundwater quality assessment in parts of eastern Niger Delta, Nigeria. Environmental Geology. 1993;**22**:41-46

[29] Esu EO, Okereke CS, Edet AE. A regional hydrostratigrahic study of Akwa Ibom state. Global Journal of Pure and Applied Sciences. 1999;**5**:89-96

[30] Ramadan A, Edet A, Merkel B, Okereke C, Nganje T. Numerical groundwater flow modeling of the coastal plain sand aquifer, Akwa Ibom state, SE Nigeria. Journal of Water Resource and Protection. 2014;**6**:193-201

[31] Goni IB. Estimating groundwater recharge in the Nigerian sector of

the Chad Basin using chloride data. In: Adelana SMA, editor. Applied Groundwater Studies in Africa. London: CRC Group; 2008

[32] Maduabuchi C, Maloszewski P, Stichler W, Eduvie M. Preliminary interpretation of environmental isotope data in the Chad Basin aquifers. In: NE Nigeria International Symposium on Isotope Hydrology and Integrated Water Resources Management, 19-23 May, Vienna, Austria. 2003

[33] Edmunds WM, Fellman E, Goni IB. Lakes, groundwater and paleohydrology in the Sahel of NE Nigeria: Evidence from hydrogeochemistry. Journal of the Geological Society of London. 1999;**156**:345-355

[34] British Geological Survey (BGS). Groundwater Quality: Nigeria. Lagos, Nigeria: British Geological Survey (BGS); 2003. p. 8

[35] United Nations Environment Programme (UNEP). Environmental Assessment of Ogoniland. Chicago, USA: UNEP Report; 2011. Available from: https://www.unep.org/resources/assessment/environmental-assessmentogoniland-report

[36] Ekwere SJ, Ukpong EE. Geochemistry of saline water in Ogoja, Cross River state of Nigeria. Journal of Mining and Geology. 1994;**30**(1):11-15

[37] Uma KO. The brine fields of the Benue Basin, Nigeria: A comparative study of geomorphic, tectonic and hydrochemical properties. Journal of African Earth Sciences. 1998;**26**(2):261-275

[38] Tijani MN. Hydrochemical and stable isotopes compositions of saline groundwaters in the Benue Basin, Nigeria. In: Adelana S, Mac Donald A, editors. Applied Groundwater Studies in Africa. Vol. 13. London, England: IAH Selected papers on Hydrogeology; 2008. pp. 352-369

[39] Sajad MM, Rahim S, Tahir T. Chemistry of groundwater of Rawalpindi/Islamabad. In: Proceedings of Water Engineering and Development Centre (WEDC), "Sanitation & Water for All", 24th Conference, Islamabad, Pakistan. 1998

[40] Matthes G. Effect of man's activities on groundwater quality. In: Hydrological Science. U.K: ITDGS Publishing; 1976

[41] International Occupational Safety and Health Information Centre. Basics of chemical safety. In: International Labour Organization Conference. Geneva, Switzerland: International Occupational Safety and Health Information Centre; 1999

[42] Nganje TN, Edet AE, Ekwere SJ. Concentrations of heavy metals and hydrocarbons in groundwater near petrol stations and mechanic workshops in Calabar Metropolis, southeastern Nigeria. Environmental Earth Sciences. 2010;**61**(2):271-279

[43] United Nations Environment Programme (UNEP). Lead Pollution and Poisoning Crisis: Environmental Emergency Response Mission to Zamfara State, Nigeria. Geneva, Switzerland: UNEP Report; 2010. Available from: https://www.unep.org/resources/assessment/lead-pollution-and-poisoning-crisis-environmental-emergency-response-mission-zamfara

[44] Mgbenu CN, Egbueri JC. The hydrogeochemical signatures, quality indices and health risk assessment of water resources in Umunya district, Southeast Nigeria. Applied Water

Science. 2019;**9**:22. DOI: 10.1007/s13201-019-0900-5

[45] Barzegar R, Moghaddam AA, Soltani S, et al. Natural and anthropogenic origins of selected trace elements in the surface waters of Tabriz area, Iran. Environmental Earth Science. 2019;**78**:254. DOI: 10.1007/s12665-019-8250-z

[46] Orubu CO, Omotor DG. Oil spillages and the Niger Delta development commission (NDDC) intervention policy in Nigeria. Research Journal of Environmental Sciences. 2011;**5**(4):327-340

[47] UNICEF, United Nations Children's Education Fund. A future global agenda for children: The links with sanitation, hygiene, water and environment. In: A UNICEF Publication on Water, Environment, Sanitation and Hygiene Water Front Issue. Vol. 13. New York, US: UNICEF; 1999. pp. 12-15

[48] Adebo BA, Adetoyinbo AA. Assessment of groundwater quality in unconsolidated sedimentary coastal aquifer in Lagos state, Nigeria. Scientific Research and Essay. 2009;**4**(4):314-319

[49] Ahmed MS, Tanko AI. Cogitation on hydrocarbon contaminants in shallow groundwater around Kaduna refining and petrochemical company, Nigeria. Ambient Science. 2019;**6**(1):42-49. DOI: 10.21276/amb:.2019.06.1.aa01. Available from: https://www.caves.res.in/

[50] Ahmed MS, Tanko AI, Eduvie MO, Ahmed M. Assessment of groundwater vulnerability in Kaduna Metropolis, Northwest Nigeria. Journal of Geoscience and Environmental Protection. 2019;**5**(6):2170428

[51] Ahmed MS, Tanko AI, Badamasi MM, Abdulhamid A. Groundwater vulnerability and risk assessment in Kaduna Metropolis, Northwest Nigeria. In: Witkowski AJ, Jakobczyk-Karpiers S, Czekaj J, Grabala D, editors. Groundwater Vulnerability and Pollution Risk Assessment, International Association of Hydrogeologist Selected Papers (24). London: CRC Press; 2020

[52] Nriagu JO, Udofia EA, Ekong I, Ebuk G. Health risks associated with oil pollution in the Niger Delta, Nigeria. International Journal of Environmental Research and Public Health. 2016;**13**(3):346

[53] Edet AE. Groundwater quality assessment in parts of eastern Niger Delta, Nigeria. Environmental Geology. 2011;**22**:41-46

[54] Dibal HU, Schoeneich K, Garba I, Lar UA, Bala EA. Overview of fluoride distribution in major aquifer units of northern Nigeria. Health. 2012;**4**(12):25292, 8 pages. DOI: 10.4236/health.2012.412189

[55] Dibal HU. Fluoride concentration in some parts of northern Nigeria [thesis]. Zaria, Nigeria: Department of Geology, Ahmadu Bello University; 2012

[56] Lar UA, Dibal HU, Daspan R, Jaryum SW. Fluoride occurrence in the surface and groundwater of Fobur area of Jos east LGA of plateau states. Journal of Environmental Sciences II. 2007;**2**:99-105

[57] Izeze E. Geology and fluoride concentration in the groundwaters of dass area [thesis]. Nigeria: Department of Geology and Mining, University of Jos; 2005